Computers: Classical, Quantum and Others

Authored By

Sergey P. Suprun

Laboratory of Heterostructure Physics and Technology
Institute of Semiconductor Physics
Siberian Branch of Russian Academy of Sciences
Novosibirsk
Russia

&

Anatoly P. Suprun

Laboratory for Cognitive Researches
Institute for System Analysis
Russian Academy of Sciences
Moscow
Russia

CONTENTS

FOREWORD

After 100 years of its progress, quantum mechanics still continues to amaze anyone who tries to understand it not restricting himself or herself to mere experimental facts and mathematical machinery. It is appropriate here to mention Richard Feynman, an outstanding physicist, who said that everybody uses the apparatus of quantum mechanics without understanding quantum mechanics. We are already accustomed to that there are efficient algorithms for computing properties of quantum systems without entering into their meaning. The interest to these problems has recently revived in connection with a heavy coverage and expectations associated with a quantum computer. The terms such as mixed states, teleportation, quantum algorithms, and others are becoming familiar not only in the scientific literature but in other fields as well. However, it is difficult to speak about the issues that lack understanding even at a professional level.

So far, there have been many attempts to interpret experimental facts from the field of quantum mechanics based on both corpuscular and wave descriptions, involving a pilot wave, and even parallel universes. All these attempts were made because the models intended to logically solve the conflict at the level of the already established notions without interfering with the geneses of these notions. This monograph, written by a psychologist and a physicist, analyzes the process of how our concepts of reality have been formed at the preconscious and conscious levels. The well-known problems in physics are used here as a touchstone for testing the capabilities of the proposed method of psychosemantic analysis. Despite that several postulates of the authors are disputable, the overall monograph deserves attention at least because it is the first attempt to puzzle out the psychological foundation of our physical knowledge. Presumably, the hypotheses proposed by the authors will motivate other researchers to look at the problems in quantum mechanics from another standpoint and notice the aspects yet beyond our field of vision. This monograph will be useful to a wide range of specialists, students, and everybody who is interested in the modern problems in science.

Prof. Viktor Ovsyuk,
Institute of Semiconductor Physics
Siberian Branch, Russian Academy of Sciences
Novosibirsk
Russia

PREFACE

The subject of classical physics is description of the changes in physical properties of objects in a space–time continuum. Consideration of exclusively object–object interactions held out a hope of completely excluding an individual observer (and, correspondingly, a subjective component) from the paradigm of natural science. However, the special theory of relativity, formulating the space–time transformations during the description of objects by observers in different frames of reference, implicitly introduces an observer into physics. Actually, the laws for transformation of reality from the standpoints of different frames of reference—those of an observer and the associated system of reference—were introduced. Thus, even the classical physics failed to avoid completely the participation of a person, an observer.

Remaining within the frame of an object-based approach to description of reality and regarding this as the only method possible, we come to the paradoxes of quantum mechanics. We assume that to introduce this field of knowledge, currently being a formal tool, into the range of our comprehension, it is necessary to admit the following postulates:

- The presence of an observer in any scientific theory is a natural fact, since it is the observer who is a "receiver" of signals and an "interpreter" of information;

- There exist evolutionarily established algorithms for assemblage of sensations (the first signal system) into object concepts realized in a certain mental space. This subjective component is also objective and should be taken into account when interpreting our concepts of reality;

- Over a long-term social development, the second signal system has acquired a semiotic object-based method for reinterpreting the reality in our consciousness relying not only on the reality per se "beyond sensations" but rather on its mirror image on a certain "mental map" that procreates our space–time concept of the world; and

- Such analytical description is not the only possible method for simulating reality. Another one is a systems approach, implying the absence of partitioning within the limits of a system and, correspondingly, the absence of space–time relationships. (In particular, the integral physical system "emitter–barrier–screen" displays interference properties, while any attempts to part it into objects immediately lead to paradoxes.)

This monograph is an attempt of consistent substantiation of the described standpoint by the example of quantum physics. The monograph comprises introduction and two parts and is written as a free discussion of the problems in question. The first part briefs the main principles underlying the operation of classical and quantum computers. The main attention there is focused on experimental facts, their interpretation, and analysis of the paradoxes in quantum mechanics.

The title of this monograph requires a separate explanation. The research in the field of informatics and formalization of the operation principles of classical and quantum computers are a mirror image of our own algorithms of consciousness. Actually, the mental processes of coding and logical reasoning have been modeled over an extended time period. Be it a conscious level or not, one more reality—a virtual one—has been created step by step within our reality with the help of computing technique. Similar to the Russian *matryoshkas,* the dolls nested into one another, these realities in part repeated themselves. Most likely, this route in science is not accidental. It looks most promising when following it with a goal in mind. This does not imply a reduction of the activity of a man of sense to functioning of a certain automaton. However, following this way, we can try to formalize the describable things, which is considered in the first part of this monograph, as well as the characters that distinguish a man from an automaton, which will be described in the second part, now in press.

All this also influenced the style of the monograph, making it multilayered. Presumably, this (as well as the content itself, extending beyond the traditional scope) requires certain efforts from a reader. The main thing for the authors was to consistently rationalize the initial postulates as much as it was possible, and we would be grateful for any criticism related to their essence.

Sergey P. Suprun, PhD

Laboratory of Heterostructure Physics and Technology
Institute of Semiconductor Physics
Siberian Branch of Russian Academy of Sciences
Novosibirsk
Russia

Anatoly P. Suprun, PhD

Laboratory for Cognitive Researches
Institute for System Analysis
Russian Academy of Sciences
Moscow
Russia

List of Contributing Authors

Sergey P. Suprun, PhD

Laboratory of Heterostructure Physics and Technology
Institute of Semiconductor Physics
Siberian Branch of Russian Academy of Sciences
Novosibirsk
Russia

Anatoly P. Suprun, PhD

Laboratory for Cognitive Researches
Institute for System Analysis
Russian Academy of Sciences
Moscow
Russia

2

CHAPTER 1

Introduction

Sergey P. Suprun[1]* and Anatoly P. Suprun[2]

[1]*Laboratory of Heterostructure Physics and Technology, Institute of Semiconductor Physics Siberian Branch of Russian Academy of Sciences, Novosibirsk, Russia and* [2]*Laboratory for Cognitive Researches, Institute for System Analysis, Russian Academy of Sciences, Moscow, Russia*

Abstract: So far, there have been many attempts to interpret experimental facts from the field of quantum mechanics but all these attempts were made because the models intended to logically resolve the contradictions at the level of the already established notions without interfering with the geneses of these notions. Here the grounds for the possible solution of quantum mechanics problems are given using the psychosemantic approach.

Any human cognition begins with contemplations, passes on to concepts and terminates in ideas.

I. Kant, «The critique of pure reason»

Keywords: Quantum Mechanics Problems, Psychosemantics, Mental Map, Semiotics.

The present book stemmed out of a series of random circumstances, and yet, it is the outcome of multi-year activities. Both authors (psychologist and physicist) have been teaching different subjects in their specialties. The conventional statement like «all problems have been solved» pronounced in the classrooms caused the feeling of dissatisfaction. First, it did not correspond to the reality and, second, it deprived students of the opportunity for creative perception of the material. For the most part, present-day courses of studies train further generations of experts who are unable to critically estimate their knowledge. Probably, this is one of the reasons why a number of points which are thought to be «inconvenient» became taboo in the scientific circles.

Teaching quantum mechanics is especially unrewarding job. It cannot be explained as there is no understanding of it as it is [1]. One can only offer a fair statement of the problems, which first requires their fair comprehension. In this way we had to turn to the very outsets - the origin and nature of human cognition, and herein lies the first cross-point of physics and psychology. Historically any mentioning of the subject, mind and «freedom of will» has always been excluded from natural sciences. Is it possible to guarantee the objectivity of our knowledge within the range of the object paradigm (in the Plato-Kantian sense)? Quantum mechanics was the first to have asked «inconvenient questions» regarding the role of the subject in a seemingly objective theory, and outlined the overlapping of the two sciences in the most problematic field. It seems there has finally appeared some hope to understand the role of consciousness in our concept of reality.

Any scientific theory or a model is the formalization of our world outlook which includes the way of its perception and semiotic representation (codification) of our knowledge. Perception patterns have been prescribed by nature that subjected *Homo sapiens* species to natural selection. From thereon, creativity begins. For some reasons, the need for communication among individuals of this species became predominant, which triggered the origin of language - a sign form of communication. Sign is already a higher degree of abstraction from direct perception. Its emergence means the formation of algorithms for the analysis of elementary sensations, pointing out the most significant ones and synthesising them into images defined as objects. Besides, a sign should possess some content to be more or less unequivocally

Address correspondence to Sergey P. Suprun: Laboratory of Heterostructure Physics and Technology, Institute of Semiconductor Physics, Siberian Branch of Russian Academy of Sciences, Novosibirsk, Russia; E-mail: serg.suprun44@gmail.com

understood by all or by the majority. It requires unification of both ways of data processing and the content that corresponds to the sign. The only way to achieve this is the exchange of information as a purposely coded signal with a common codification system. Such coordination of codes and meanings led to the emergence and maintenance of common mental space. Thus, the process of education and upbringing became an obligatory function of social life. In this way we have upgraded our development so that initial details become indistinct as separate elements in integral circuits. To understand the function of these devices, it is necessary to turn back - to the times of resistance, diodes and vessels.

It seems that the above-mentioned does not directly deal with physics. However, how did the field of knowledge that is most successful in its application, ended up in a difficult situation when dealing with explanation of reality (its semantic interpretation) behind experimental data? First of all, the points of interference, reduction of wave function, measurements [2] are meant. In the present it has become impossible to resolve these problems within the physical theory only, and it is due to this fact that the content of this book is interdisciplinary. The selection of content material was aimed at the clearest demonstration of existing problems, on one hand, and the consideration of possible approaches to their solution, on the other. Moreover, it was meant to show not only the possibility of solving paradoxes but also to consider the legality of the proposed methods. It was a rather challenging task, yet we made an attempt at it.

Paradoxical as it may seem, but references to cognition and consciousness became a tendency in physics when resolving paradoxes of quantum mechanics, whereas in psychology they prefer to avoid these contemporary fundamental issues. Such works as «The Emperor's New Mind», «Shades of the Mind» by Roger Penrose, «God reason» by Paul Davies and others can be a vivid example of this.

The inevitability of overlapping of «physical» and «psychical», as physicists believe, lies in the field of investigating consciousness:

> ... a chain of measurements cannot interrupt in any physical system. However, we know the point where there is only one alternative, *i.e.* the measurement is already over. It is consciousness. Hence, an endless chain leads to consciousness. That is why the «problem of measurement» is not, in our standpoint, a physical problem and... may be solved (if it is possible in general) only including such a concept as consciousness.

Michael Mensky [3]

The theory of reason will help science explain its own origin.

Gerald Edelman [4] [1]

> Neither information science, nor neurology can imitate the intuitive power of the mind. There should be a kind of more profound theory that explains quantum paradoxes and confuses subjective elements. Finally, this theory should be a shelter to subjectivism, not being subjective itself. Such a theory will need some inducing naturalism. It is to make sense...

Roger Penrose [4]

> One and the same elements are used to create both internal (psychological) and external (outer) world... Subject and object are united. One cannot say that the barrier between them is destroyed as a result of the achievements in physical sciences, as there is no such a barrier.

E. Schrödinger [4]

---

[1] In the book cite preprint by Princeton University, 1983, Robert G. Jahn and Brenda J. Dunne «Collected Thoughts on the Role of Consciousness in the Physical Representation of Reality".

Consciousness and matter are different aspects of one and the same reality...

C. Weizsaecker [4]

Thus, there developed the understanding of the necessity to reconsider the base of our knowledge in physics. The question «what to begin with?» - remained open. A new approach is to have such a degree of universality and integrity that would allow us to explain the paradoxes in different sciences, considering their integration level nowadays.

Taking into account the fact that basic experiments in quantum mechanics have already moved out of the purely physical field to the philisophical-methodological level that tackles the basic principles of reality organisation, including Consciousness, we think it necessary to turn to the foundations of our knowledge of the world again. How true is it that intrinsically consistent axioms underlie our ancient «self-evident» prejudices of the ways of obtaining information about the world? Is the version of what we call «conscious perception» really an adequate «reflection» of the thing that we believe to be external or part of surrounding reality? It is even more topical, as modern psychological studies demonstrate, there is a deep layer of «subliminal» phenomena which are closer to physical reality than reasoned psychic interpretations that follow [5-7]. It is also obvious that any research method and presentation of information have their limits. How do traditional forms of knowledge presentation affect our reality outlook? Do not they mislead us and, if they do, then in what aspects? We would restrain from approaching the analysis of the discussed problems from the philosophical standpoint, yet we would like to focus our efforts primarily on their experimental and logical bases, no matter how unusual they would seem to be at first sight. At the same time, considering different alternative approaches, we aimed at the level of modern physics in which the most significant data had been experimentally obtained and confirmed. Unfortunately, we face one of the most difficult problems on this way - the linguistic problem. Considering different variants of reality interpretation, we inevitably change the meanings and sense of basic concepts including those of physics. We did not have the audacity enoughto create a new language. That is why we had to specify the understanding of this or that term in different paradigms. This fact should be specially considered, since we doubt whether we have always managed to achieve accuracy in treating these issues. Unfortunately, it took a while for the authors of this book to realise these details: many things have been rather elusive due to their «self-evidence».

As the objective and interdisciplinary outlook at the existing paradoxes in quantum mechanics and consciousness was our main goal, we set out with a «naive» understanding of the terms trying not to adhere to any existing paradigm. Therefore we would kindly request advocates of existing scientific trends not to rush into accusing us of «reductionism» or plagiarism. As Carl Valentine, a Bavarian comic, once said: «Everything has been said before, but not by everyone». Many issues have been raised by different authors within more than 2000 years; so it is impossible to trace all the ideas and priorities. The authors do not intend to do it on any account. Their task is seen in the logical accordance of experimental facts and quite reliable formalisations with our mental representations of reality. A wide and unbiased discussion of the problems and finding possible solutions are our priorities in this contribution. Unfortunately, using the language means applying some terms in the most conventional sense, which considerably restricts our possibilities of presenting the problems from a different angle, as it creates a great number of subconscious settings.

As cognition is the only conscious way in our representation of «external reality», we should focus very attentively on its mental modelling process. The only way out, as we believe, is the analysis of the perception process itself and algorithms of semiotic encoding of reality by human mind. Herein, there is a possibility to reconsider a wide circle of scientific matters applying semiotic and psychosemantic methods to analysis of ways of constructing object reality. Mathematical modelling of the subject's «mental map» is the most effective instrument of analysis that allows us to point out objective regularities of collection, storage and transformation of a set of characteristics that we define as object. It is obvious that, at this stage, the subject begins to play a considerable role in the defining of what we call «objective» reality and it must find its proper place in the universal world outlook including its «physical» constituent.

Making up semantic spaces is a way of mathematical construction of mental map as a system modelling «objective reality» by the type of mathematical structures (arbitrary multitudes with certain relations). As it will further be demonstrated, in the process of formalisation of semantic analysis method, a number of correlations were discovered, that coincide with basic physical laws, yet they are far beyond the purely physical field. Thus it indirectly proves the correctness of such an approach to scientific reality description that embraces a new role of the subject in scientific cognition. Consequently, the research subject of semantic analysis could be formulated as follows: in which way does the subject's world reflection affect our knowledge about it? Semantic analysis presupposes the existence of common non-specific regularities within any science related to «topological» characteristics of the mental map, the characteristics being determined by peculiarities of reality representation on the map.

Despite the «apology» of the subject in science, the principle of subjectivity remains declarative even in the sphere of humanities, and the scientific theory still «successfully» confines the subject to the object, in spite of the «non-classical» statement that objective reality may be reasonable only when correlated with the subject. Semantic description naturally introduces the subject into any theory and allows us to resolve a number of paradoxes connected with the role of «observer» (*e.g.* in special relativity theory and quantum mechanics).

Along with this, the methodology of semantic analysis turned to be oriented not as much to the investigation of *mechanical processes in the classical paradigm* as to studying *creative processes* (development, evolution, creativity) which cannot be the subject of classical science because they cannot be determined in a formal system, as development itself presents non-finite (non-terminal) content. In principle, it is impossible to integrally describe the infinite content using the classical finite (after Goedel) theory. However, it does not exclude the possibility of recurrent studies of evolutionary processes. In the paradigm of semantic analysis, they can be revealed through a consecutive hierarchy of sense-forming systems of different levels in which the sense of the preceding system is determined by the «evolutionary need» of the following one. The problem of creativity and the possibility of its theoretical investigation will be discussed in the next volume.

Consideration of creative processes evidently presupposes applying the principle of integrity which we correlate with the principle of subjectness. The idea of the subject as an attribute of the world unity, was introduced by us in the concept of subjectness as a hierarchical system of the subject's sense redefinition. Such an approach, with logical inevitability, requires rejection of the classical causality principle and our preconceptions of spatial-temporal relations, as well as reconsidering the categories of the subjective and the objective, subjective and objective organisation of psychic reality.

In principle, semiotic and psychological bases of the semantic analysis allow us to pass on to semantic modelling of «psychic reality», in which universal conservation laws, as well as principles of relativity and complementarity, are observed. In a particular case, these general regularities can be confined to physical, psychological, biological, *etc.* laws. It is still not clear how far this formalisation can be continued and how successful it will be. However, there is a possibility of convergence of different approaches to the understanding of universal Reality that does not exclude either subject or object.

Some words on the structure of this book. It consists of four parts and is written in the form of lecture abstracts with comments containing free discussion of the problems. In Part II the basic principles of classical computer functioning are considered. Part III is devoted to the analysis of paradoxes in quantum mechanics and the possibility of making the quantum computer. It is mainly focused on the formulation of experimental facts and their interpretation. The understanding of these things is necessary to picture the limits of the possible. As it was sharply observed in the book [2]: «the Ptolemaic system is bad not because the Sun orbits the Earth, but because, using it, we cannot launch a satellite». Correct reality representation allows us to see the way out where there used to be a deadlock.

The next volume consider the points of consciousness and creativity formulated in the same key as the first two. They are quite an independent contribution.

Probably, it is necessary to explain where the title of the book originated from. Research in the field of informatics, formalisation of the principles of classical and quantum computers functioning are the reflection of our own algorithms of consciousness. In fact, modelling of coding processes and logical reasoning has been in progress for a long time. Whether consciously or not, one more virtual reality has been gradually created through computing machines within our reality. They replicated themselves as Russian matryoshkas placed inside each other. Such a way is, obviously, not random in science. It is very promising if you pursue this path by keeping focus on the goal. *Homo sapiens* activities are not confined to the functioning of an automatic device. However, following this way, one can try to formalise the things that can be described, which the first two parts of the book are devoted to, and the next two parts dwell on the differences between a human and a machine.

All the above-mentioned affected the style of the book's material which ended up to be «multi-level». Probably, this (also the content itself, which does not conform to usual stylistic settings) requires certain efforts in reading and often causes emotional aversion. Successive argumentation of initial standpoints was our main objective as far as it was possible. Any comments or notes are always welcome.

ACKNOWLEDGEMENTS

Many people have helped us during the work on this monograph, and we would like to thank them for their assistance and support as well as for the time they spent on reading the manuscript and their helpful advice. Undoubtedly, the authors only are responsible for any flaws and shortcomings in this monograph.

First and foremost, we are grateful for fruitful discussions on physical problems to Vladimir Shumsky, Viktor Ovsyuk, Yury Tsidulko, and Igor Neizvestny, as well as many others, including our students, who by their curiosity made us resolve the problems that we believed to have no answers. We want to thank Viktor Petrenko, Natal'ya Yanova, and Konstantin Nosov for discussing the issues related to psychology.

Our special appreciation for our wives, both Galinas, for correcting the text, comparison of the Russian and English versions, and their patience and understanding during our work on the manuscript. We thank very much Sergei (the son of S.S.) for his technical assistance in preparing the manuscript.

We would like to thank the Bentham Science Publishers, especially Asma Ahmed, manager, who gave us the possibility to realise our intention. We also acknowledge the work of our translators - Elena Gileva, Galina Chirikova and Alexander Zhuravlev - who had to work with the terminology of sciences so distant from one another.

This work was supported by the Russian Foundation for Basic Research (project no. 10-06-00192).

REFERENCES

[1]   R. Feynman, *The character of physical law*. Cox and Wyman LTD, London, 1965.
[2]   G. Greenstein, A.G. Zajonc, *The Quantum Challenge. Modern Research on the Foundations of Quantum Mechanics*. Jones and Bartlett Publishers, Inc., 2006. (Intellekt, Dolgoprulny, 2008 (in Russian))
[3]   M.B. Mensky, *Quantum measurements and decoherence*. Physmathlit publ., Moscow, 2001.
[4]   V.V. Nalimov, *In search of other meanings (in Russian)*. Progress, Moscow, 1993.
[5]   A.R. Luria, *Fundamental neuropsychology (in Russian)*. Publ. of the Moscow University, Moscow, 1973.
[6]   A.R. Luria, *Language and Mind (in Russian)*. Feniks, Rostov-na-Donu, 1998.
[7]   S.A. Nadirashvili, *Psychological nature of perception at the disposition theory (in Russian)*. Publ. of Uznadze Instit. of Psychology, Moscow, 1976.

CHAPTER 2

Classical Computer

Sergey P. Suprun[1*] and Anatoly P. Suprun[2]

[1]*Laboratory of Heterostructure Physics and Technology, Institute of Semiconductor Physics, Siberian Branch of Russian Academy of Sciences, Novosibirsk, Russia and* [2]*Laboratory for Cognitive Researches, Institute for System Analysis, Russian Academy of Sciences, Moscow, Russia*

Abstract: The classical physics is considered in the paradigm of an object-based representation of reality and, correspondingly, the classical computer is considered as a logical device implemented in an object-based manner (made of elements - objects) that functions according to a space-time pattern following a selected algorithm.

He who controls probability, rules the world

Midnight thoughts

Keywords: Information, Shannon Entropy, Entropy in Physics, Boolean Functions, Contactless Logical Elements, Basic ECM Blocks, Cryptography.

HISTORY OF THE ORIGIN OF COMPUTING MACHINES

Computing machines occurred in connection with the necessity of searching precise solutions of difficult tasks for a limited time span. Such devices as electronic modelling machines, in which changes of physical values proceeding in real processes are replaced by voltaic changes in the electric circuit with specially matched parameters, were developed first. The base of these systems functioning is that various phenomena that we observe can be described using the same mathematical equations and simulated on analogous devices. Different periodical processes can be used as an example: pendular oscillations, current changes in the electric contour, spreading of waves in matter, *etc*. The advantage of such modelling machines is their high tasks solution velocity; they can fix the dynamics of very complicated processes which are described by a big number of non-linear differential equations. Such systems found their special application in the solution of automatic control and regulation tasks in the real time format. Their drawbacks consist in the restricted precision of calculations and, what is more important, scarce universality of their application, as each of such machines is developed to solve one concrete task.

Optical computing systems are one more example of computing devices functioning according to optical laws, and the computation process is confined to the transformation of optical signals. Videoinformation processing is their basic application field. Optical computing machines can carry out such operations as Fresnel, Fourier and Laplace integral transformations and realise these operations in parallel with a two-dimensional array; as a consequence of this, the velocity of operations implementations is determined by the times of information in- and output. They have optical memory devices of big volumes and high rapidity. As optical signals are two-dimensional (signal intensity is spatially distributed), operations are realised in parallel in such devices, and they determines their high efficiency.

And, finally, the classical electron computer has become an indispensible device and is determined much in the image of the contemporary society due to its high rapidity and universality. Further we will dwell upon its construction and the way it functions. Computer is the most wonderful device created by man. It is difficult to imagine all fields of its application, but its basic purpose is information processing and its representation in a convenient form of perception.

---

*Address correspondence to Sergey P. Suprun: Laboratory of Heterostructure Physics and Technology, Institute of Semiconductor Physics, Siberian Branch of Russian Academy of Sciences, Novosibirsk, Russia; E-mail: serg.suprun44@gmail.com

In his book «The fabric of reality» D. Deutsch [1], author of the first quantum computer algorithm, formulated such an idea as: developing a modern powerful classical computer that would lead to the occurrence of the virtual 'habitat'. True, at present, it is possible to generate sound, visual image, intact sensations and even to synthesise smells using a set of certain initial substances from a special device [1]. Thus, a space suit, in which a person can enter the virtual world, not distinct from our reality, has already been created. D. Deutsch really states the thing that it is possible to differentiate medium-environment and reality as one cannot immitate weightlessness. It is so if not to affect certain brain centres which cause the feeling of weightlessness. Well, if there is a direct access to the cortex and a way to disturb its certain centres, there is no need for such a space suit, as human body proper is a 'virtual reality' space suit. Irritating different sensory and motor centres of the brain (posterior portions of the cerebral cortex, cingulate gyrus and some others), it is possible to cause any motion reactions, sensory and emotional reactions in an individual. Under such an approach to the creation of virtual reality, it principally remains indistict from ours. «If there is no difference», as it is announced in advertising, is it worth producing new entities? It is really an interesting question and not that simple either.

In this connection there comes an idea that classical computer simulates some part of consciosness activities: the final functional stage - that of with coded data. The way a program is realised will be considered further on. The relevant moment is that computer processes information, *i.e.* there are values, at the in- and output, to which certain objects, relations or processes are compared to.

SOME DETAILS ABOUT INFORMATION

Information, Shannon entropy, information unit, the concept of entropy in physics, Maxwell's demon, two-dimensional switch-over element, energy cost of information.

> *It is only for God that any signal is information*
>
> *(if He really needs it at all)*
>
> ***Midnight thoughts***

The term «information» is unusually volumous. Man is its final consumer; he accepts and interprets the content. However, only quantitative aspect of information is considered in natural sciences. This aspect is connected with coding and decoding processes (reading and registration) which are accompanied by a change of a system - «material carrier» of information. Variants of coding, transfer of channel parameters optimisation, physical limits for transfer velocity, *etc.* are analysed in detail in different fields of informatics, mentioning the final information consumer being evaded. Obviously, such a situation is connected with the thing that, in natural sciences, using the term «subject» (observer is a less categoric variant) gives a hint at the subjectivity of information content. There follows a suspicion as for the reality proper studied by science. It is possible to treat the problem in another way; so, when formulating the material in a traditional form, we will focus on the points which are thought paradoxical in scientific papers. They are the key moments of our non-understanding and, hence, the most interesting for further analysis.

There is some incoherence in the absence of «subject» in the natural sciences paradigm. Classical physics investigates the world of macro-objects and is focused on the description of object-object (physical) interactions. It is believed that it is possible to abstract from the participance of an experimentator in the experiment. It means that the interference level, *i.e.* the process of measurements that is proper or receiving a signal can be made small. Herein there is an implicit logical contradiciton between the unlimited measurements precision and the possibility of an unlimited small impact for the object. We either permit an immeasurably low influence on the object in measurements - which means the existence of precision limit - or one has to admit that the process of obtaining information from known objects leads to the disturbance of

---

[1]Obviously, this set of initial ingredients can be not big based on the number of human olfactory receptors. They are 7.

a measured object's state (here we do not consider the possibility of measurements with unlimited precision for a limited time span). Moreover, there are no grounds to extrapolate such statements into the field of micro-objects and, it means, to exclude an observer from the process of obtaining information. Not concerning the results obtained in quantum mechanics so far, it is possible to note that the «observer» already appears in the classical theory, such as Special Relativity Theory (SRT). It is from his (observer's) standpoint that observations and measurements of certain properties of focused-on objects are carried out in one's countdown system. The existing limitation for information or a signal transfer (transport) velocity leads to the relativity (or are they still subjectivities?) of the concept simultaneity of events proceeding in different countdown systems (*i.e.* for different observers).

> • *The observer (and not one, they even differ in names - Alice, Bob, etc.) entered quantum mechanics as experimentators that carry out measurements. Already one can already not do without him in quantum mechanics. If we want to be successful, it is necessary to acknowledge «observer» in whole physics. It is only on this condition that a transition from quantum to classical mechanics, as its special case based on the same ground, is possible.*
>
> *On the other hand, not everything is clear in the definition of «observer». If a certain observer has a set of definite physical characteristics, it makes his immediate transfers from one countdown system to the other, «on his own wish», problematic. If it is possible to consider physical processes from different countdown systems, either different observers can do it or an additional procedure is required to compare their data. There is a feeling that we implicitly try to be someone who can consider this world through different windows that are different observers with their own countdown systems. In this sense, the countdown system is such a localised spatial-temporal window. The remarks made are the introduction of the problem. Let us ask any questions, no matter how unusual they would seem at first sight, but we would not like to restrict the understanding of problems with formed conventionalities and settings.*

Consider the quantitative aspect of information. In his times, Shannon introduced the notion of information entropy which characterises possible states of a system depending on the probability of their realisation. Let a system realise its *i*-state with probability p_i. For instance, a coin, if it is not false [2], when tossed, will have its «tail-up» state with probability 1/2. Then, according to Shannon, full entropy of the system can be written as $H = -\sum_i p_i \log_2 p_i$, where the logarithm is on the base 2. The information to be extracted when observing the outcome of the event will be $I = H_2 - H_1$, equal to the difference of entropy before and after the event. It is necessary to note that the value of $0 \cdot \log_2 0 = 0$ is determined here (also $\log_2 1 = 0$), so:

$$H_1 = \frac{1}{2} \cdot \log_2 \frac{1}{2} + \frac{1}{2} \cdot \log_2 \frac{1}{2} = -1$$

$$H_2 = 1 \cdot \log_2 1 + 0 \cdot \log_2 0 = 0$$

$$I = H_2 - H_1 = 1$$

For a system having two states with equal realisation probability, the quantity of extracted information is equal to 1, and this unit was rendered the name of BIT.

Everything seems simple at first sight if not to ask other questions. For example, what is probability [3] ? This notion is supposed to be intuitively clear: it is the occurrence of frequency of some outcoming result in

---

[2] *i.e.* one side is heavier than the other, or both sides are «tail-up»'.

[3] Actually, probability is the measure of possibility which is a category of the future, but it is «measured»' and, hence, it has its «presence»' in a form in the present. If you think it over, it will sound a bit strange.

a number of certain multitude of events. Probability is the norm per unit, the probability of an impossible event being equal to zero. What an impossible event is - is also intuitive. If you lived in the times of Fermat, then the probability of getting to Rome from Moscow would be equal to zero, which would not be in Fermi times. It is worth paying attention to the paradoxes of probability theory [2, 3]. It has turned out so that the language of mathematics is very volumous in its content and brief in writing, without extra-details which describes certain relations of our perception. In the theory of probability, conditions of events observation are not considered, a countdown system is not introduced; we are not interested in the points where the left or right sides are and how the top differs from the bottom. But here different nuances occur. For example, if there are two indistinguishable coins but no outlined direction, then only two variants of their location in one plane are possible: either onto one side with the same sides or with different ones (physicists would say «singlet state»), and it is equally probable for not interacting objects. The number of variants increases if an outlined direction is introduced. Besides, it is possible to consider the case of distinguishable numbered objects. Probabilities of events observation become different in these conditions. In V. Boss book «Counter-examples and paradoxes» [3], there is a detailed specification of the difficulties of logical description and understanding seeming contradictions, provided at the observation details omission of described random events.

- *Passing on to the probability distribution functions allows us to exclude the space of elementary events* Ω *- which we, actually, used when solving tasks (doing sums) considering the variants of possible outcomes - from direct analysis. The real line or its sub-multitude becomes the space of a random value. However, a more profound level, if it is, remains beyond vision. For instance, is cabbage yield a random value? It obviously is, but it is affected by so many factors that we can only guess the presence of deep* Ω *. God throws bones from that side, and here we observe the result - the random value proper. Even in the simplest case of tossing the coin, the «crest - lattice» space is an agreggated illusion. Real* Ω *should be searched at the other level where some more is known about the fabric of the Universe [2]. Obviously, there is a principle of physical difference between classical and quantum probability which mathematics simply ignores. If the first one depends on our non-wish to consider all «life trifles» (and here is the subjective element again!), the second one does not depend on us. We cannot comprehend its meaning because it is vividly determined at the «other level» (not necessarily physical). We find ourselves in the situation of a linguist analysing texts in an unknown language. The sequence of letters is not random, and we cannot do any other thing but to estimate their occurrence probability (anyway at the quantitative level of information estimates).*

The concept «information» became most widely spread in the description field of computing systems functioning on the base of binary elements. Although there is a number of other tasks which can be considered within the probability of different outcomes and the possible volume of information obtained during tests.

- *An interesting basis for the case of peculiarity of the probability is also the thing that it cannot be presented graphically on the orthonormed basis, e.g. for two events. If we describe the result of tossing the coin, then the vector that reflects the oucome of the event will be on line* $y + x = 1$ *in the first quarter of the circle, the vector's length being not constant and not equal to 1 but the extreme points: on both sides of a coin with heads or tails Fig. (2.1) i.e. it is possible to intorduce the orthonormed basis, e.g. in the Hilbert space, for probability amplitude, and this will be dwelled upon further on.*

If we consider the concept of information physically, it is necessary to emphasise the thing that the process of its fixation is unbreakably connected with it. Information can be read, transmitted, received or used in any other way if it is fixed during some physical process. In other words, any information has its physical representation [4] and «energy» cost. The famous proverb about the thing that «there is no information for free» can be proved scientifically.

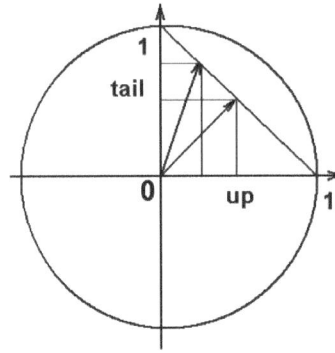

Figure 2.1. Probability of a «false» coin.

In fact, when speaking about information, a physicist, in contrast to a mathematician, always implies a way of its transmission and fixation. In this case it can be connected with a change of a certain information carrier. It is clear that the process of recording (fixation) and preservation, as any other physical process, requires energy consumption. It is this peculiarity of information that is most brightly illustrated by the paradox proposed by Maxwell where the demon breaks the second law of thermodynamics getting the output by observation and sorting of gas molecules without any loss of energy.

The gist of the paradox consists of the fact that any closed system tends to achieve its equilibrium, which is most probable, *i.e.* it is realized by the biggest number of possible ways - events possible for development. This obvious statement is introduced in physics as a postulate on the impossibility to decrease a closed sytem's entropy.

> • *Herewith the concept of «closed system» proper is idialisation of the case when some part of the world is believed to be not interacting with the surroundings. It is useless to create such a system. One can simulate the trajectory of molecules in two vessels under completely identical initial conditions only with one difference: movement of mass 1 g over 1 cm at the distance of light year from the Earth. It turns out that they will completely diverge in the time less than 1 sec [5]. In connection with the thing that it is impossible to screen the effect of gravitation, it seems obvious that speculations about a closed system are just an approximation.*

In physics, entropy, unlike Shannon definition, is equal to natural logarithm N of possible system's microstates: $S = k \cdot lnN$, where k - Boltzman constant. Nevertheless, these notions are connected.

Any thermodynamical equilibrium is a dynamical equilibrium: the system fluctuates near the most probable state. It means that, at some moments of time, it will have a less probable state. For instance, a number of molecules in a vessel, if mentally divided into two equal parts, may be bigger in one half at some moment of time than in the other. «Demon» (it is a device that traces a system's state and fixes the number of molecules in each half of the vessel) is present in one of paradoxical variants (Fig. (**2.2**)). Then, getting down the shutter at the required moment of time, using a mobile pistol in the channel that connects both vessel halves, it can make it move to the right side obtaining the output without any additional energy source. In this case, kinetic energy of gas molecules is consumed to carry out the work, breaking the second law of thermodynamics: for example, it is impossible to transform the heat taken from a body into work (output) as a whole without any other changes in the nature (or it is impossible to develop a perpetum mobile of the second kind). The solution of this paradox lies in the consideration of the demon's physical activity proper which has to fix a system's changes (molecules position in a vessel) as information was provided about it. The end of the demon's cycle is accompanied by the erasure of this information which is equivalent to the obtained output [6]. Generally speaking, the solution of the paradox is conditioned by the introduction of energy expenses for some process connected with information retrieval about a system's state and, finally, with activities of the demon proper.

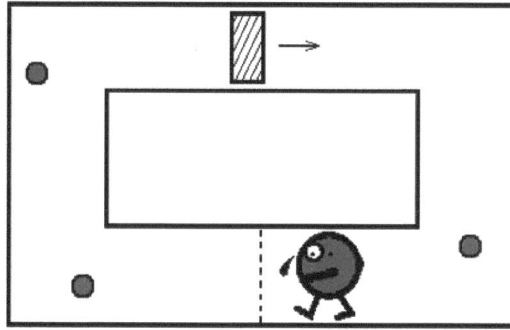

Figure 2.2. Maxwell Demon.

In the R. Landauer contribution [7], there is a detailed analysis of the functioning of a two-positional switch-over device and how the energy is consumed to change its state. It is obvious that, to transfer from one state to the other, *e.g.* from «0» to «1», it is necessary to overcome a potential barrier between them. The device has to function without friction and losses, excessive kinetic energy is unpermittable, as, in this case, it will be necessary to introduce the process of faltering that suppresses the system's fluctuations to transfer to the stable equilibrium. It is clear that such a device will be slow-functioning, and here we have to choose one of the two things - either the time spent or energy. Moreover, to exclude the self-arbitrary transition from one position to the other, the value of potential barrier is to be higher than that of the system's thermal energy equal to kT (k - Boltzmann constant, T - absolute temperature). Landauer showed that the minimal cost of infromation bit will be $k \cdot \ln 2$ (it is necessary to consider the temperature at which a system functions) that connects the concept of Shannon and physical entropy. It is clear that real computing systems spend more orders of energy per bit.

- *Physically, is there any difference between a machinary information process and that of the bioloigical system? Physics is believed to be one and the same everywhere. It means that a change of a state at the molecular level is also accompanied by the energy change. True, this process is very economical compared to energy consumption of classical computers. Consciousness is not supposed to be some computing algorithm here. This aspect was worked out in detail in R. Penrose [8, 9] contributions. But, at least, any logical activity is accompanied by some information process connected with a change of the physical state in our nervous system [9].*

  Formally, entropy can be introduced only in a closed system as equilibrium is achieved only in this condition. However, temperature cannot be determined in this case, as it is introduced as a process of setting thermal equilibrium with a thermostat, and interaction contradicts to closedness [10], the difference between the mathematical and physical concept of entropy and information becoming clearer. In mathematics, extracted information is connected with the description of this or that event's outcome from a definite set, i.e. «past-in-the-present» is described after the choice. Quantum mechanics analyses the «future-in-the-past» situation describing it before the choice. Physics can consider the process of obtaining information and analysing the interaction of an observer with the investigated classical or quantum system. However, such an approach has not been vividly realised so far.

Thus, to obtain information, energy consumption is required and, after that, it is possible to take to its quantitative processing. What is called information proper and is its source, is not considered in sciences. It implies that there exists a certain stage at which signal processing proceeds and the value (meaning) is rendered. It is the most relevant point, and it is after this that information may be used in communication. If we evade the answer, then there will be a situation of indefiniteness and alleged objectivity. However, quantum mechanics has already demonstrated the possibility of immediate state transmission, but not that of information (Part II). Velocity limitation exists for information processes and there is a feeling that there exists an element of «subjectivity» in it.

LOGIC OF CLASSICAL COMPUTER

> *Logical operations with multitudes, Boolean functions, record and arithmetic calculations in the binary system, (electron-computing machine) ECM model and examples of micro-operations*

Elelements of the Boolean functions theory (or functions of algebra of logic) which find their applications in ECM designs, programming and other fields of cybernetics are put down in this paragraph.

- *Besides, logic is an instrument using which we make our judgements. Logical non-contradiction is the basic acceptability condition of the proposed theory in the explanation of experimental facts. There, simultaneously, may be several models which highlight the same problems from different angles. Then the choice is determined by the convenience of use, simplicity of computations or other factors.*

First, let us introduce the basic concepts for Boolean operations with multitudes: integration, cross and addition. *Integration* of A and B multitudes $(A \cup B)$ is the multitude of all elements each of these belongs to, at least, one of multitudes A or B; elements belonging to both mutitudes A and B are in integration $(A \cup B)$ only one time. *Cross* of $(A \cap B)$ is the mutitude that consists of all elements simultaneously belonging to both multitude A and B. *Addition* of multitude A is the multitude of all elelements $(x \in E)$ belonging to universal multitude E and not belonging to multitude A $(x \notin A)$. Venn diagrams that explain Boolean operations with multitudes are presented in Fig. (**2.3**).

- *Function $f(x_1, x_2, ..., x_n)$ is said to be Boolean if all its variables $(x_1, x_2, ..., x_n)$ and function f proper have values from two-element multitude $E_2 = \{0,1\}$. A totality of all possible n-local ordered sets of variables values called binary or simply sets serves as domain of the Boolean function from n-variables.*

   Boolean function from n-variables is called n-local. The domain of n-local Boolean function consists of 2^n sets of variables values which include all possible combinations 0 and 1.

#1	2	...	...	...	$n-1$	n
0	0	0	...	...	0	0
1	0	0	...	...	0	0
0	1	0	...	...	0	0
...	...	...	...	...	...	...
1	1	1	...	...	1	0
1	1	1			1	1

   The finity of domain and range of values of an arbitrary Boolean function allows us to set the function using tables.

   Let us consider binary sets of variables values as a registration of some integers in the binary system so that set $A = (a_1, a_2, ..., a_n)$ is identified with registration

$$a_1 \times 2^{n-1} + a_2 \times 2^{n-2} ... + a_{n-1} \times 2^1 + a_n \times 2^0 .$$

Thus, for $n= 4$, Fig. 5 will be written with the following set of 0101:

$$(0 \times 2^3 + 1 \times 2^2 + 0 \times 2^1 + 1 \times 2^0 = 5)_.$$

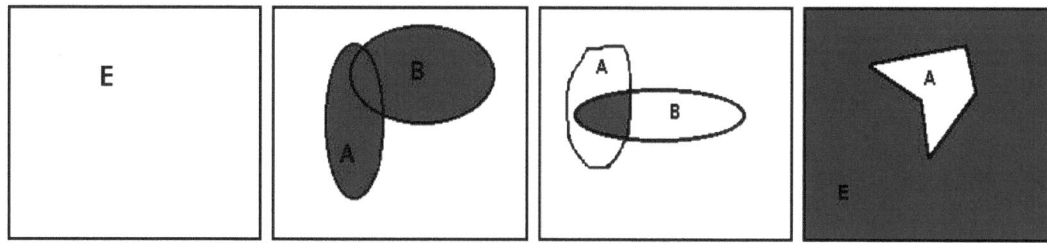

Figure 2.3. Venn diagrams explaining Boolean operations over multitudes.

Modern ECM are complexes of devices supplied with advanced mathematical provision systems which include languages of programming and translators from them, operation systems, dispatchers, various service programs, also systems of standard and typal sub-programs. All these means can be realised both as a program and schematically. The process of working out the systems, just as ECM design, is most complicated and laborious. Composition and ways of developing mathematical provision systems are considerably connected with the structure of a certain machine and should be ready by the moment of its fabrication. Therefore, working out the systems of mathematical provision should be arranged simultaneously with ECM design. The last thing is also true of quantum computer. Further on, it will become clear that one has to solve the tasks analogous to those of classical computer.

The concept of abstract ECM model, which is the composition of two automats - operation and control - belongs to the number of basic ones in the theory of algorithmic algebras. So, the function of each ECM, in spans between addressing a new command, can be described as interaction of these two automats [11]. Output signals of the automatic control system are identified with micro-operations realised by the automatic operation system.

- *For example the following components belong to the set of micro-operations and logical conditions:*

  $\mathbf{e}$ -   identical transformation;

  $\mathbf{s}_{ij}$ -   addition of the i-register content to that of j (i-register code being unchanged);

  $\mathbf{p}_i$ -   addition of the unit to the i-register content;

  $\mathbf{l}_i$ -   one order shift to senior bits on the i-register;

  $\mathbf{r}_i$ -   one order shift to junior bit on the i-register;

  $\mathbf{o}_i$ -   erasure (setting the zero code) of the i-register;

  α_i -   condition of equality to zero of the i-register content;

  β_i -   condition of equality to zero of all bits content with non-positive numbers (including the zero bit) on the i-register.

  To examplify this, let us consider the transformation of a linear part of the operation device microprogram that consists of one abstract binary register; and let us think that all the code bits set on the register, beginning with zero and higher, are the integer part of the corresponding (binary) number, negative bits being its fractional part. Then it is easy to obtain the ratio $\mathbf{p}^2 \cdot \mathbf{r} = \mathbf{r} \cdot \mathbf{p}$ from the content of the corresponding binary number.

True, operator $\mathbf{p}^2$ (consecutively twice realised addition of one to the register content) in the left part transforms the code of number x set on the register into code $x+2$. Further using the shift operation for $\mathbf{r}$ at one bit to junior bits on the register will transform this code into $1/2(x+2)$. For instance:

$$(1000.0000) \Rightarrow \mathbf{p}^2 (1000.0000) \Rightarrow (1010.0000) \Rightarrow \mathbf{r}(1010.0000) \Rightarrow (0101.0000)$$

8   10   5   *Analogous*

transformations of $r \cdot p$ transform the code into $(1+x/2)$.

Developed microprograms of different transformations allow us to realise all mathematical operations using only two logical functions, *e.g.* addition and complementation. Thus, to prove the possibility of developing a computer, including quantum computer, it is necessary to suggest the process and to describe the function of elements which carry out the indicated logical operations.

The main thing to be focused on is that, at the present, a detailed mathematical language - in which it is possible to formulate a process of logical speculation as a certain algorithm - has been developed. Realisation of some algorithm as a program is a sequence of comands, which means its representation as a process extended in time. This remark is relevant because the presence of spatial-temporal terminology denotes the classical way of Reality description. Besides, it was experimentally established that part of neurons, when disturbed, function in the regime like «zero-one» transmitting a sequence of uniform impulses in neuron nets that consist of axons and neurons [12-14]. It corresponds to the functioning of the automat involving Boolean functions algebra in binary coding, which is suggestive of the possibility of simulating certain consciousness activities by classical computer.

CONTACTLESS LOGICAL ELEMENTS

Contactless logical elements, table of states, basic ECM blocks, «Moore's law»

The first computing devices were mechanisms assembled from cogwheels, just like clocks. Those were quite complicated devices despite their allegedly primitive element base. Thus, in the XVIII century, Swiss foremen - farther and son, Henri and Pierre Droz - made «mechanical people» that wrote and drew, *e.g.* the portraits of Ludwigs XV and XVI, also those of Marie Antoinette. Later, computing machines were assembled on the base of other elements, but their essence did not change from that: all of them were automats object realisations with a fixed (stable) or variable program. This statement is also fair of modern classical computers.

Logical devices functioning on the base of semiconductor elements are called contactless logical elements in schemotechnics, - unlike relay automatic machines on the basis of which realisation of Boolean function is also possible. For example, logical element «OR», the table of which is presented lower, can be assembled on two diodes (Fig. (**2.4**)).

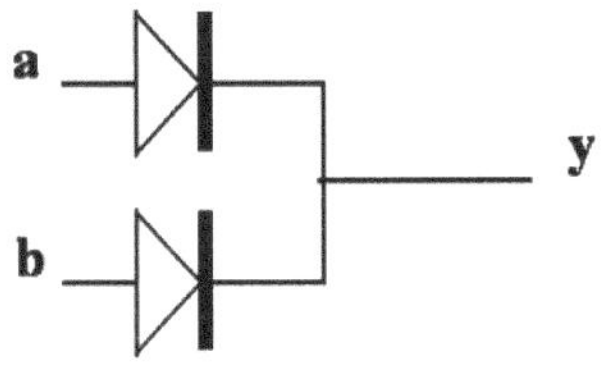

a	b	+
0	0	0
1	0	1
0	1	1
1	1	1

Figure 2.4. Structure and matrix of logical element «OR».

In the signals equal to «0», inputs **a** and **b** will also have «0» in output **y**; in signals equal to «1», in any input, the output will also be «1». Diodes, unlike the pevious case, are switched in here in the reverse direction. If there is «0» in any of the inputs or in both of them, then the current flows from the positive power supply pole, passes through diodes whose resistance is much lower than R, and the whole voltage decreases in the resistor.

In the output, **y** will be «0». In the positive shift equal to «1» in both **a** and **b** inputs, both diodes become locked; the current does not flow through them, and it wll be «1» in the output (Fig. (**2.5**).

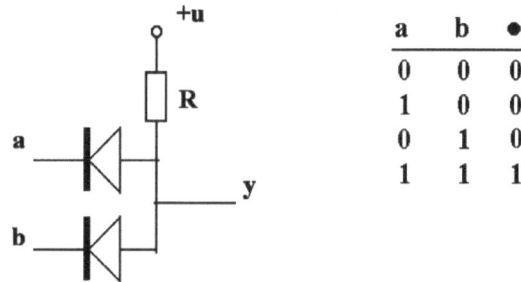

a	b	●
0	0	0
1	0	0
0	1	0
1	1	1

Figure 2.5. Structure and matrix of logical element «AND».

Logical element «NOT» is an amplifier on the semiconductor triode functioning in the key regime whose output has the voltage equal to «1» at «0» in the output; and, on the contrary - «0» at «1» in the output. The table of the logical element «NOT» significance is shown in the following Fig. (**2.6**).

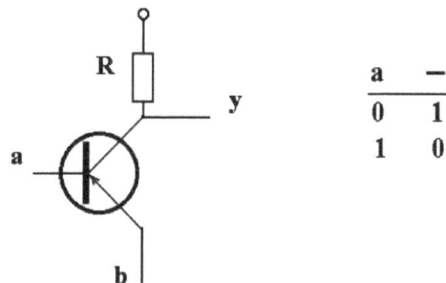

a	–
0	1
1	0

Figure 2.6. Structure and matrix of states for logical element «NOT».

At present there are microcircuits fabricated to realise certain logical operations which are the element base of all automatic devices and computing machines. The major blocks which any ECM consists of are the memory device which comprises a non-volatile memory device, immediate memory device and external memory block. Besides, an ECM has an arithmetic logical device, control, data in-output, data representation and control desk.

- *The aim of the arithmetic logical device is the realisation of the basic arithmetic and logical transformations of information. All the information necessary for work is stored in the memory device. It includes the block of non-volatile memory device for the storage of information which does not change during work and is preserved after any switch-off. Immediate memory device is for the temporary preservation of intermediary computations and programs storage. External memory is for long-term information storage and it has a huge volume. Information representation device is for visualisation of current data, i.e. display. In-output device is for information in- and output. The control device carries out temporal concordance of separate blocks functioning, its functions include sending commands and data among internal registers, control of computation process, interaction of external devices. Arithmetic logical device, which forms a processor with the control device, is the central part of ECM, and it subdues all the rest blocks and units serving this automatic computation device. Registers are the main information carriers inside the processor; the number of bits in each register determines the EMC production output as a whole. Such a structure of CM was proposed by John von Neumann. ECM of the last generations possesses a considerably more developed architecture already having «intellectual interface». However, the principle of functioning has not changed [15].*

Since the XIX century attempts have been made to create a universal computing machine capable of solving tasks formulated and recorded in a certain language as a sequence of codes - programs. It has led to the advancement of our outlook in such scientific fields as formal logic, informatics, cybernetics and a number of others. For this purpose, one had to analyse and simulate the process of reasoning - a logical chain of the intial reasoning transformation (input) to the conclusion (output). It is believed that, within a theory with a set system of axioms, any supposition or a theoreme is a tautology. Thus, the proof is the transformation of the basis which it is carried out in. One can use such a metaphor. Imagine you observe one facet of a 3D body, but you know the length of sides, volume and angles between links. Transformation of the co-ordinate system or viewpoints on this object will make its shape obvious for you, and this is the sense of gist's proof. An ECM can do it if basis of transformation equations and a required «aspect angle» are known.

Development of the theory of algorithms analysis was an important stage in making the computer. In 1936, Alan Turing proposed a model of the automatic computing machine - Turing machine - to formalise the concept of algorithm. It allowed us to divide them into groups with a carried out estimate of algorithmic complexity. Thus, a group of algorithms corresponding to class NP. Class NP (Engl.: «non-deterministic polynomial») is a multitude of algorithms whose work time is determined by the volume of input data and it polynomially depends on their length. Actually, the tasks, which can be described with some formal language and solved using the classical computer, were determined. But it was not clear how and what to solve other tasks with. Something was missing for a full understanding of the problem in the field of algorithmic complexity.

Besides the above-mentioned ECM blocks, cooling system is one of the most necessary blocks. As it is known, the classical computer dissipates a considerable amount of heat during its operation. It occurs because recording and erasure of the information bit in logical elements are an irreversible process connected with a transfer of charge carriers in electron circuits and are accompanied by energy dissipation. In connection with electron elements miniaturisation, their density highly increased per unit of volume and that led to higher density of dissipated energy.

Already in 1965, Gordon Moore had noticed the thing that electron circuit elements density became doubled for about two years. Such a tendency in electronics development was rendered the name of «Moore's law». At present, this regularity tends to remain in its limit. Complexities are connected with the homogeneity in characteristics of basic elements of computing circuits - transistors, in particular, due to thinning of the sub-shutter dielectric, channel length reduction and a number of other things. Thus, technically, the development of classical computer functioning in the consequtive operation regime comes to an end. On this very scheme, according to the realised program, at every temporal tact of computer functioning, there is a transfer of the signal from one element to the other during the successive realisation process of a number of commands according to the program being carried out. To accelerate functioning, the division of the program into parts, which can be realised in parallel with the use of several processors, is used. However, to change the situation cardinally, we need some other new ways of further computing devices development.

Now let us wind it all up. We have an object-realised (constructed on the base of elements - objects) computing machine - classical computer. Along with this, on the «spatial-temporal» scheme, a sequence of commands (program) - generally performing a logical transformation - is realised. It is possible to have a principally different element base to which another computer and functional principle will be adequate to, but it will be considered in the next part of this book.

CRYPTOGRAPHY

> *Prime numbers, comparisons and their properties, biggest common divisor, Eiler function, open-key cryptosystem.*

This paragraph is to clear out the situation that caused interest in the problem of developing quantum computer for the last period of time. It also acquaints us with new characters - Alice and Bob - who further study quantum cryptography. Skimming the content of the paragraph will allow you to have an idea about this point if there is no necessity in getting absorbed. Herein, we will briefly analyse the basic problems of classical cryptography which has become widely spread now due to the use of computing devices. Most elementary descriptions of information coding techniques and ways of key transfer for its read-out will be given.

Occurrence of personal computers (PCs) and Internet has led to ubiquitous use of different coding systems in information transmission. In the exchange of secret information, the transmission of key, using which message decoding is made, is the most vulnerable thing. It turns out that there are ways of information transmission through open channels which provide its secrecy (**Appendix A**).

Let us think it over, how we can code the transferred message. For instance, each letter of the alphabet can be rendered a number corresponding to its position in the alphabet, just as it is shown in the first and second lines of the following Table (**2.1**) (addition on module 26).

Table 2.1: Sample of coding.

Letter	A	B	C	D	E	...	...	Y	Z
number	00	01	02	03	04	...	...	25	26
key	15	04	28	13	14	...	...	16	10
code	15	05	04	16	18	...	...	15	10

If there is a set of random numbers called key (third line of the table) which can be compared to this line of numbers, then, by means of the simplest operation, - by addition of letter number to key number (fourth line of the table) - the code is obtained. It is not difficult to restore this message if there is a key. Using a long key, whose length corresponds to that of the message text, is most reliable. In this case one can evade the code replication. Moreover, one-time use of the same key is obligatory. However, the problem of its safe transmission remains open. To resolve this task, cryptosystems without key transmission and those with open-key [16] (**Appendix A**) were developed. In scientific literature; such a description was made on the information transmission between two points or persons marked as A and B. Further, names Alice and Bob were used in the description of these experiments.

Those to whom it may concern may analyse the principle of open-key cryptosystem's operation after the book of Simon Singh «Book of codes» [17]. And we think the conclusion that, even for small prime numbers, when intercepting codes, it will require certain computing resources to decode a message, to be important. In serious instituions, prime numbers of 100-200 signs long are used. To break these codes, you will need an ample amount of time - hundreds of millions of years to be more precise. Therefore, in 1994, when Peter Shor published his quantum algorithm of factorisation (decomposition of number into simple co-factors) that requires polynomially growing time related to the value of number, he attracted universal attention. This algorithm provided decoding messages for several minutes. The focus on the problem of developing quantum computer was transferred from the theoretical into practical aspect, and it is clear why.

- *One can presuppose that the Nature, developing such a nanotechnological combine - which the biological cell is - could not help using such an exciting computing power: the time required for the solution of factorisation task using quantum computer is related to the time of classical computer functioning as* $1:(10^{13}-10^{14})$ *! These are totally different possibilities and another world.*

CONCLUSIONS

Modern applied algebra has its roots in pure algebra and symbolic logic. The latter, in its turn, originated from introspective psychology. For instance, Boole - 150 years later than Leibniz - began his classical tract «Investigation of the Laws of Thought» with such an explanation of the «character and aim of this contribution»:

> The aim of the suggested tract is investigation of the most important mindful acts with which consideration is realised; giving them the expression in the symbolic countdown language and, on this base, establishing (statement) of the science of logic and developing its method; then making this method the base of the general one for the application of mathematical probability study; and, finally, collecting some probable indications regarding the nature and structure of human mind out of different elements of truth.

Now it is important for us that classical computer simulates this very causality-determined part of our mind which we refer to as consciousness (deliberate). Note that classical computer, actually - just as all classical physics - allows us:

> ...to describe events development in the nature as something causal -extended in space and time (or in the relativistic space-time) and, hence, to create models which are clear and precise for a physicist; but modern quantum physics prohibits any representations of this type and makes them totally impossible. It permits only theories that underlie by purely abstract formulae and it doubts the idea about the causal process of atomic and corpuscular phenomena; it admits only probability laws; it considers these laws as the ones which are primary and make the finite cognizable Reality; it does not permit their explanation as a result of causal development which takes place at a more profound level of the physical world...

> L. Debroglie in the foreword to the book by D. Bohm

> «Causality and Chance in Modern Physics»

However, if not to pull out forced «psychical Reality» from that of *per se*, then quantum laws, having «primary reality», may relate to the unconscious constituent of our mind. It is not excluded that our consciousness (or its type) is not «primary», but it has quite a complicated structure. «Classical algorithms» of «primary Reality» presentation to our consciousness may, principally, lead to a whole number of paradoxical perceptions and logical contradictions if we set this very reflection (or representation) type as initial Reality. Therefore, it is impossible to answer the question about the origin of aroused contradictions right on the spot, without a detailed analysis.

First, it is necessary to find out clearly how conscious object representations are realised and how adequate they are to this very «primary Reality» and, if they are not, then what distortions they add to it. In this aspect, it is especially interesting to trace how the idea of causality arises in general (to be more profound - that of object spatial-temporal Reality representation) and how adequate it is to the truth (generally, how legitimate demarcation of the so-called «physical» and «psychical» reality is).

REFERENCES

[1]   D. Deutsch, *The Fabric of Reality*. Allen Lane, The Penguuin Press, 1997.
[2]   V. Boss, *Lectures in mathematics (in Russian), V.4*. KomKniga, 2004.
[3]   V. Boss, *Lectures in mathematics (in Russian), V.12*. KomKniga, 2008.
[4]   A. Steane, Quantum Computating. Quant-ph/9708022.
[5]   E. Borel, *Introduction geometrique a physique*. Paris: Gauthier-Villars, 1912.
[6]   C. M. Caves, "Quantitave Limits on the Ability of a Maxwell Demon to Extract Work from Heart," *Phys. Rev. Lett.*, vol. 64, pp. 2111-2114, 1990.

[7]   R. Landauer, "Irreversibility and heat generation in the computing process," *IBM. J. Res. Develop.,* vol. 5, pp. 183-191, 1961.

[8]   R. Penrose, *The Emperor's New Mind: Concerning Computers, Minds, and the Laws of Physics.* Oxford University Press, Oxford, 1989.

[9]   R. Penrose, *Shadows of the Mind: An Approach to the Missing Science of Consciousness.* Oxford University Press, Oxford, 1992.

[10]   B. H. Lavenda, *Statistical physics.* A Wiley-Interscience Publication, John Wiley and Sons, Inc. Oxford, 1991.

[11]   V.M. Glushkov, G.E. Tseitlin, E.L Yushchenko, *Algebra, languages, programing (in Russian).* Naukova dumka, Kiev, 1974.

[12]   J. McCarty, and C. Shannon. *eds. Automata Studies. Annals of Mathematical Studies, 34,* Princeton University Press, Princeton, 1956.

[13]   M.L. Minsky, *Computation: Finite and Infinite Machines.* Engle-wood Cliffs, N.J., Prentice-Hall, 1967.

[14]   N. Wiener, *Cybernetics*, 2nd ed. N.Y.: John Wiley: Cambridge, Mass.: MIT Press, 1961.

[15]   *Informatics for all (in Russian).* In the Collect. Personal computers, Nauka publ., Moscow, 1987.

[16]   V.I. Nechaev, *Elements of cryptography (in Russian).* Vysshaya shkola, Moscow, 1999.

[17]   Simon Singh, *The Code Book by Simon Singh.* Fourth Estate Limited, 2000.

CHAPTER 3

Quantum Computer

Sergey P. Suprun[1*] and Anatoly P. Suprun[2]

[1]Laboratory of Heterostructure Physics and Technology, Institute of Semiconductor Physics Siberian Branch of Russian Academy of Sciences, Novosibirsk, Russia and [2]Laboratory for Cognitive Researches, Institute for System Analysis, Russian Academy of Sciences, Moscow, Russia

Abstract: The problem of observer and system of reference in physics is considered as well as an object-object and an object-subject approach to a physical description of reality. The evolutionarily formed methods underlying data processing of the first signal system and the algorithms for assemblage of sensations into object concepts and representation of reality in a space-time continuum are discussed. The methods for constructing a mental space as a tool for sign-based description of reality during information exchange are considered. The problems in understanding the quantum phenomena as the reality not subject to an object-based description are briefed, and the need in a system and integral concept as a concept adequate to reality is substantiated. The principles underlying the operation of a quantum computer, its distinctions from a classical computer, and the difficulties in presentation of the issues in question using an object-based language are discussed.

We do not understand many things not because our concepts are weak, but because these things do not belong to the circle of our understanding.

Kozma Prutkov

Keywords: Semiotic limitations, mental map, linguistic modeling, object-based, objective, subject-based, subjective, Subject in physics, observer in physics, wave function reduction, experimental bases of quantum mechanics, probability amplitude, rules to deal with probability amplitudes, object-based modeling of reality, metric of mental space, definition of qubit, entangled pairs, Von Neumann projectors, collapse of wave function, psychological and physical times, determinism and evolution, consciousness and reality, criterion of Einstein-Podolsky-Rosen Reality, Bell inequalities, mechanistic process and development, evolution and problem of choice, representation of phenomenon in space- time and Hilbert space, space-time and spectral windows, physical reality and selection system, qubit quantum gates, quantum algorithms, dense coding, teleportation, quantum computer block-scheme.

INTRODUCTION

Passing on to the point of making up quantum computer, as a historical reference, it is necessary to note that the idea about the possibility of using quantum systems as information carriers was first formulated by Yu. Manin [1]. Later, R. Feyman dwelled on and grounded the supposition about creating the computer functioning with principally different elements according to other non-classical laws of physics [2]. These activities have become more intensive since the XX century up to the present. Actually, it is a deeper level Reality simulation.

Further on, we will base on the statement that information, as a whole, can not be considered separately from the subject as its final addressee and consumer. Taking into account also that physical laws are indifferent to our division of matter into «living» and «non-living», the understanding of such a device as quantum computer allows us to consider the activity of our own consciousness in a new way. Finally, the reality «out of us» and «in us» should logically cluster to make these worlds integrated. It is hardly possible to make up the theory of «all-in-the-world» without this thing.

*****Address correspondence to Sergey P. Suprun:** Laboratory of Heterostructure Physics and Technology, Institute of Semiconductor Physics, Siberian Branch of Russian Academy of Sciences, Novosibirsk, Russia; E-mail: serg.suprun44@gmail.com

ABOUT THE ORIGIN OF OUR KNOWLEDGE

Semiotic limitations for description of reality, signaling character of perception, mental map, linguistic modeling of reality, problem of "limit" concepts, object-based and objective, subject-based and subjective.

There should be a kind of more profound theory that explains quantum paradoxes and makes subjective elements feel bewildred. Finally, this theory should be a shelter to subjectivism, which it is not. Such a theory will need some inducing naturalism. It is to have reason.

R. Penrose

Starting from a leading special case, we find a general solution assisted with superposition of special cases.

D. Pouya

To understand the paradoxes of quantum mechanics, it is necessary to consider the problem from one side, as all the currently used purely physical ways of explanation have not led to any relevant results. So, let us begin from the outset and ask ourselves a question: how is our knowledge about the world formulated?

All sciences study different aspects of *uniform Reality*. Thus, it is not suprising that once their ways are to cross. Some traits of this «cross» are noticed even now in those difficulties and problems which «flow over» from the fields of methodology and general science into those of experimental technology and formal logic. These problems are most obvious in such sciences as physics, psychology, biology, mathematics and linguistics. Let us try to lead the paradoxes, that appear in paradigms of these approaches to Reality, to some general bias and to show their similarity. It is more convenient to begin with the problems of linguistics because language, as a semiotic (sign) system, is the universal common scientific means of Reality description in any science, and its problems include all scientific trends.

Linguistics or, in a broader sense, semiotics, as a science of signs, has its own axiomatics and logic (rules of «output»). Disturbing them will inevitably cause errors in semiotic Reality description, including mathematical, as the latter is also a language.

The principles of «conditionality» and *«differentiation of meanings»* of any sign are the basic ones in semiotics. It means that sign does not have any direct meaning, but it points it out. Its semantic content is conditioned and, because of this, it is the base of the second signal system (after I.P. Pavlov). Besides, a sign's content is determined by its distinctive potential. It is possible to agree and formally differentiate, using a sign, arbitrary aspects of reality which are relevant for orientation and survival under general conditions of our existence. Conditional pointing out of one Reality aspect means its opposition to the other and that, finally, determines the outcome of a sign's content.

Sign «conditionality» implies that along with the «objective» (human-independent) component, it also includes the «subjective», an individual-dependent component. Let us explain it with examples.

Signs appear only *in the necessity to realise certain individuals communications* in a socium. Sign differentiates direct and indirect factors that are important for our existence, sign meanings being very different in different languages. For instance, for Chukchas, snow may be an important means of spatial orientation as there are traces of predominant winds in it. On a nuber of traits, it may be suggestive of the presence of lichens (genus *Cladonia*) under its cover - deer feed. These two meanings differentiate various features of one «object» and are indicated with different signs. From Chukchas, the need for survival required more then twenty words to denote the notion «snow». It is obvious that, for an urban dweller, having cardinally different needs, these meanings are not topical. «Object» (meaningful) world discretion turns to be completely different for urban people.

Obtaining any information about reality begins with sensations (feelings) - signal system I after I.P. Pavlov. Sensation is *a psychic process* that reflects *the state* (disturbance) of a subject. The following aspects of sensations are worth noting:

1.   The signal (marker) that reflects some reality aspect but not the reality proper. For example, disturbance caused by red colour does not mean that a photon of certain energy or electromagnetic wave of certain frequency has colour.

2.   Sensation (feeling) is realised when it is differentiated by a subject out of the general psychic state and opposed to it as something separate, compared to one's «body» or «environment».

3.   Initially, sensations cannot be connected with the process of obtaining *information* in the form of a sign, as this process implies coding stages, transmission over the connection channel and decoding. However, in the other «end of wire» nobody codes anything and nobody is going to send us a message. Actually, we have to interpret signals on our own and to ascribe certain meanings to them. It is only after this stage that they become signs and can be considered as information having sense for us. Our adaptation towards the thing that we conditionally call «environment» (external medium) is the only criterion of adequateness of such interpretation. Conditions of transition from the state of unsatisfied need to that of psychological comfort are operationalised by the subject as *«external means»* of needs satisfaction (objects, actions, conditioned reflexes, *etc.*).

4.   Signal system I differs from the second one (discursus, language) in that it is addressed to you personally and cannot be a communication means with other «subjects» in this form.

It is possible to point out the following ranges of sensations:

•   *Exteroceptive sensations* (vision, hearing, smelling, taste, intact feelings) which basically do not depend on our arbitrary will and they define what we usually connect with the «external surrounding».

•   *Introceptive sensations* (hunger, thirst, pain and others) also do not depend on our arbitrary will but are referred to the «internal medium» (body).

•   *Pro-preoceptive sensations* (movement, motor sensations, spatial orientation) are those that depend on our will and directly belong to the «subject» or body.

Having its history of one millenium, the debate about the «source of feelings» has not led to any one opinion. Most of scholars believe that, on «that side» of feelings, there exists a special «world of things» independent from the subject (as it is in Platon's ideas). In this «subjectless» world, there exist such things as tables, atoms, galaxies, mushroom pies, collections of Sheakespeare and Dostoyevsky's books *per se* which, affecting our «organs of feelings» of our body initiate nervous impulses. Further on, by means of unset mechanisms, impulses will recode into feelings (sensations) proper and will be transmitted to the «subject» through an unestablished connection channel. As the physical problem has not been solved, and its solution will not be outlined in the nearest future, we will not go into discussions about this «on-the-other-side» world. Let us only note that, we have no direct exit into this world and, therefore, its recognition is connected with an act of belief, which is a bit beyond the scientific paradigm. This point is very delicate and often painful. If we donot have to tackle the problem of belief, we have to admit that, either the world of sensations is the only reality we have or to believe that it is a specific semiotic (sign) system that translates the information to us from «outside». So far, we have no possibility to resolve the point of «outside» without taking to occultism. Thus, the basic problem becomes the question about the thing how we, based on these «signals», discrete the objective information and make up the «world of objects» or objective reality on its bias.

Let us focus one more time on the term «information». In a general case, this is an *aimful* process in which the one who sends and codes information, its transmission channel and the receiver for the one who receives, decodes and interprets it are implied. We have to use the formed terminology, but if to approach the problem in a stricter way, then we do not have any right to name the processes as signals even for the so-called «external disturbance». It is only in the result of the whole series of autonomous operations, after the correlation of external (exteroceptive) and internal (interoceptive [1]) processes and finding their «correlates», separate signals - relevant for an individual [2] - to which setting of perception and further realisation occurs. Then follows «noise» filtration of currently untopical processes, then remembering, correlation with other signals, *etc.* Next, only after that signs of the first level or feelings [3] are constructed from the pointed out signals.

To have a clearer understanding of the bases of such interpretations, let us consider the following example. Suppose we are in an isolated room in which there are various displays: sonic light indicators, arrow devices, *etc.* that reflect the «outer world». However, we do not know what they are responsible for and how they are generated. We also have «organs of control»: buttons, arms, *etc.* (motor system). However, we do not know with what and where we control. Besides, we have indicators of our «welfare» whose worsening may lead us to lethality. Can we, based on the devices (and «welfare indicators» - the so-called «inborn non-conditioned reflexes»), - nevertheless, survive in these conditions? It is obvious that we will have to compare the changes in sensor and motor systems to welfare indicators and to connect sensor complexes and motor patterns with welfare achievements in this or that field. In this way, as a consequence of analysis and generalisation, that objects of external world having gist (connection with a need) and meaning (description) for us, which we can put on the «external Reality» map, occur in our imagination.

Having admitted the *«signal»* character of feelings, we have to recognise the *model, semiotic* (sign) character of our hypothetic «subject Reality» outlook on «that side» of these feelings. Clear understanding of the thing that, object reality is, essentially, our *model* outlook on the hypothetic source of the objective (*i.e.* independent from individual mind) constituent element of our feelings. As making up of mental representations begins at *the pre-conscious level*, then, it is obvious that not only relations among objects, but also those objective algorithms on which these objects are pointed out and made, should be included in the scientific consideration. It is not a simple task if the biological sensitive formation period of a concrete function is missing. For instance, teaching adult patients, who have their cataracts removed, to focus out objects from the visual field may last for several decades.

In principle, it is possible to discrete the visible reality into constituent objects and to obtain equally effective survival models of this reality. Such distinctions are marked by linguists dealing with comparative linguistics. But predominant similarity of the world's linguistic map in many peoples makes us suspect the thing that there are some common principles, objectives for species *Homo sapiens*, of such discretion.

We will think that, as a result of some common objective (for a certain social group) and, mostly, subconscious processes, an individual - based on the objective class of feelings - creates a mental model in his or her *mental space*, which he or she believes to be the source of these feelings, *i.e. object reality* or *the world mental map*. Then the map (or its fragments) are translated in the form of signal system II or a language to other subjects; on follows checking the objectivity of these messages (coincidence of mental representations), comparance of the efficiency of different interpretations to achieve some aims [4] and recoding them into the scientific semiotic form (ideally, it is mental map reflection as a mathematical

---

[1] connected with the field of an individual's needs.

[2] *i.e.* connected with the satisfaction of certain needs and orienting an individual in the search of «external means» to compensate «internal homeostasis» aberrations (meaningful).

[3] Adaptive setting process proceeds at early ontogenesis and leads to the thing that, by the age of 10 months, children become functionally deaf for sounds absent in their native language. The map of a child's hearing develops by approximately the age of one year. Though six-month-old children can differentiate each phoneme in such different languages as Hindu and Nslakampks - an Indian language whose combination of consonants is indistinct for anyone but its informants [3].

[4] and it means their truthfulness.

model or theory). Thus, science, using verbal reality representation, investigates not the world proper, but its mental «reflection» - our outlook on it, and different branches of sciences use their own semiotic and conceptual systems for this purpose. But all of them, finally, realise *linguistic modelling* of our Reality outlook, as language is the general instrument of any phenomenon's description for all sciences.

> Taking to classical notions is... finally, the result of mankind's spiritual development... Concepts of classical physics are specified notions of our everyday life and they form the most important constitutive part of language, which is the pre-condition of all natural sciences... We should remember that, what we observe, is not the nature proper, but the one which manifests itself due to our way of problem-setting. Scientific activities in physics consist in the formulation of the questions about nature in the language we use and in the attempt to get an answer in the experiment carried out with the means at our disposal (V. Heisenberg [4]).

> • *As a consequence of the thing that science deals with secondary modelling (simulation) (if the «subject reality» hypothesis is recognised), the division of the components of scientific knowledge becomes not such a simple task. For example, we believe, that a house preserves its dimensions with no external impact. This invariability is connected with its proper and does not depend on the subject. However, when changing the distance from the object, an observer will notice its visual increase or decrease, this regularity being in accord with strict perception laws of geometrical perspective and, hence, it is also objective, though it should be referred to the category of subject but not object regularities. It is not the «law of nature», but rather a regularity connected with the way of reflecting this very «nature» onto our mental map. The seeming decrease of an object's dimensions marks the distance from it. In other cases, it is more difficult to discrete and separate these components. Further, we will show that part of the thing believed to be laws of nature is, obviously, subject regularities.*

We can see that object reality occurs as a consequence of alienating part of «sensor(ial)» reality (experience of disturbance) by the subject, which one interprets «external» regarding him- or herself. His or her own body is self-treated dubiously: on the one hand, it belongs to object reality, as it interacts with other objects and subdues to physical laws; on the other hand, - to that of subject, as he or she correlates his/her own disturbance experiences with it (body). Within physical limits, the subject can control his or her body directly and «experience» its physiological states; by the way, not only at the level of visual and aural sensations [5], which are correlated with the «external environment» by him or her, but at the level of those which are referred to one's «internal reality» [6]. As «alienation» of object reality by the subject proceeds along the line of a certain class of sensations, the body occupies an intermediate (and contradictious) position in the opposition of two types of reality in this interpretation.

Thus, any field of scientific knowledge inevitably has two «biases» as its ground: psychology and semiotics, as the semiotic way is the only one in obtaining our knowledge and its representation, and any other away of acquiring knowledge - evading signal systems I and II - is not accepted in scientific circles. Herein, it follows that axiomatics of semiotics is a real restriction of the thing that can be described within classical science. Semiotics is a means of considering anything as signs and their systems; EVERYTHING that denotes something is in its focus. The other question is: can everything be expressed in signs?

In object ontologies, there is the following system of conditions:

> Objective = Object;

> Subjective = Subject.

---

[5]Epicritical sensitivity, most closely connected with cognitive functions on H. Head's classification [6].

[6]Protopathic and ancient forms of sensitivity indivisible from emotions on H. Head's classification. Note that H. Head believed that proto- and epicritical components are present in any modality of sensations, but in a different proportion. *E.g.* vision and hearing are mostly determined by the epicritical component, introceptive sensations (feelings) - protopathic [6].

It means that anything that belongs to the world of objects, inanimate bodies, non-organic processes are believed to be objective, universally true and, thus, scientific in object ontologies. On the contrary, everything that deals with the subject in any way, should be excluded, taken out from the content of scientific knowledge. Classical physics kept to these very standpoints till the XX century. Heinsberg wrote: «Believeing in causative conditionality of all events, thought of as objective and independent from the observer, was made this way the basic postulate of new European natural sciences». For example, Einstein believed that physics, to have any value for him, should satisfy his need for evading a subject with one's arbitrariness. Nevertheless, in General Theory of Relativity (GTR) he (subject) is really God observing such a thing as completely determined Universe independent from him. In his letter to Max Born (1924) Einstein wrote that, if he had to refuse from strict causality, he would prefer «to make a shoe-maker or a croupier in a gambling house, rather than physicist» [5].

But perception of the environment is impossible without any act of «interaction» with it. Even our distant ancestors understood that if there were no «subject-environment» interaction, they he or she would not be able to perceive or be perceived. For example, they left vulnarability to their mythical characters (remember Achilles and Siegfried). *Excluding subject from the scientific paradigm means excluding also perception of the studied reality proper.*

Scholars of natural sciences determine their position as object or objective, *i.e.* the one investigating the objects that constitute Reality as entities independent from the subject - objects *per se*. In such understanding, objectivity required the exclusion of the subject [7] totally from the explanatory paradigm (and, finally, out of Reality proper). Following the same tradition, in psychology, they also tried to make their own science as *objective*, *i.e.* independent from the subject «arbitrariness», and it inevitably came accross the paradox: studying psychic *objectively*, like in natural sciences, one also has *to exclude the subject* out of its paradigm. But, in fact, it led to the exclusion of the subject of investigation. The attempt of studying the subject in this tradition inevitably and not vividly replaces it by different types of objects: *ego, individuality, personality* and others understood purely mechanistically [8]. Any axiomatically set (finite after D. Hilbert) theory is *mechanistic* and *definitely* describes its subject and its evolution in time, completely excluding «arbitrariness» of the subject. The attempt to find the way out of unequivocal predefiniteness, by means of *probable* description of the investigation of subject, does not change anything essentially. Simply *«evolution of probabilities»*, which still remains purely mechanistic, is considered as an object, as probabilties proper are calculated unequivocally. The only aspect which remains «beyond» in the theories of this type is the point of how the real choice out of possible alternatives is realised, *i.e.* the mechanism of *possibility-reality* transformation in each concrete case. As we will see further, this problem has become the corner stone in quantum physics, giving rise to a number of paradoxes - unresolved in the classical, purely object paradigm.

One should note that the attempt of psychologists to get out of the «object» mechanistic definition by means of «polyparadigmal pluralism» also leads to strange concepts allegedly made on the principles of determinism, but having a non-determined element which totally crosses out all prognostic possiblilities of the theories for which they were being developed 9. For example, in the S. Freud concept [7], personality is considered as a structure of three interacting axes: *ego, super-ego* and *id*. As for the first two, they are quite «deterministic» and, in principle, they are based on the «sublimation», «ousting» and other mechanisms; they allow us to causatively explain and predict an individual's behaviour. However, *id* is not determined. At any moment anything may «jump out from there», and it is after this that the psychoanalyst begins to develop his explanatory mythologemes. If, in the S. Freud theory, the undetermined element is «located» inside a personality, then, in behaviourism, it is «outside», as the rule that transforms stimulus into reaction is based on the impact of *external random events* not determined within this very concept («Operant conditioning» [8]).

---

[7]with his subjectivity, freedom of will and other «inconvenient» categories.

[8]This is what all psychodiagnostics is based on as a prognostic theory.

[9]If scientific knowledge does not provide us a possibility of orientation in a situation, *i.e.* to evaluate the consequences of decisions we take, then it is useless for us as scientific, anyway.

At first sight, it seems that the «subject» problem is purely psychological and it can be «painlessly» evaded in other fields of knowledge. However, it is not this way. Niels Bohr reported on that in 1930: «whereas the relativity theory reminded us of the subjective, considerably depending on the character of an observer's viewpoint for all physical phenomena, the unbreakable connection of atomic phenomena with their observation, originating from the quantum theory, using our means of survival, induces us to be as careful as in psychological problems where we are restlessly affected by the difficulty in the distinction of objective content from an observing subject» [9].

Let us note that, if we do not use the equality sign between the objective and object, the subjective and subject, then not only elements of subject reality may be subjective, but also those of the objective - e.g heat production device (caloric), ether or, in some people's opinion, superstrings, *etc*. At the same time, such «subject» disturbances (feelings) as hunger or thirst are vividly objective, and ignoring them may lead to quite predictable and sad consequences.

And, again, to the semiotic aspect of this problem. The basic function of sign is differentiation of meanings; there is no sign without it [10]. According to semiotics, meaning is set by opposition: *something is always determined by anything* other [10]. Thus, the definition of object (studied by any science) is determined through the «object-subject» opposition. So, if we even do not include the notion of «subject» in the explanatory paradigm, it is still implicitly present in it as a necessary element of defining the object and, sooner or later, we will inevitably have to consider its «presence» in our logical speculations.

Obviously, most of researchers will agree that the object way of world description is, first of all, *analytical* and «localisationistic», as it is limited by what is inevitably realised in limited semiotic units. We «break» the fabric of Reality into local elements and consider them as separate axes of the Universum. Thus, the outlook on Reality as *one whole* is the opposition of such an approach. Analytical approach is conditioned, is proved in the sense that artificially «broken» world -even for its local description - requires the introduction of various «pseudo-objects» (forces, fields, *etc.*) that partially put them together into one whole. Such a way of cognition, certainly, has the right to exist and enables us to resolve many practical tasks. However, it is not self-sufficient, as it ignores the other Reality aspect - its unity.

Pointing out some properties of the world during perception is determined by the necessity of a subject for orientation in the «object» reality to successfully satisfy topical needs beginning with the vital. Properties are peculiar conditional «parallels and meridians» of Reality. The world has no marking without the subject proper; it is as there is no co-ordinate net on the Earth's surface. We can point out or not point out some properties of the world [11], but we have to oppose the quality poles in this point-out due to the way of perception proper. For example, the world *per se* is neither ideal and nor material. However, there may be some necessity for such a distinction on some purposes. But, according to the laws of semiotics, in the conditioned definition of the *material*, we have to oppose the *ideal* to it at the very same moment. To argue on what is primary is the same as to argue about what side of the coin occurred earlier. Hence, the definition of *object* or *object Reality* is reasonable only when it is opposed to *subject* and *subject Reality* that determine each other in the opposition only [12].

The problem is the conditioned semiotic *object world* that is put on the mental map, in essence, which coincides with the thing commonly understood as being, *i.e.* what *objectively exists* and *«unconditionally» is*. But Non-being (not-being) or Nothing, for which the quantor of existence cannot be already used, is the opposition to Being. To state the thing that Non-being would mean to include it in Being, and we will have a mass of purely logical paradoxes. The same troubles occur in the semiotic making up of such concepts as Reality, Universe, Infinity and Absolute. It has been known for a long time that logical contradictions grow because such «concepts» are semiotically inexplicable. For instance, in the middle ages, there was one

---

[10]Sign is to carry at least one information bit.

[11]Pointing out some properties has its genetic and evolutionary determination.

[12]The concepts «object Reality» and «objective Reality», just as «subjective Reality» and «subject Reality», are not equivalent.

known paradox: can the *all-mighty* and *omnipotent* God create a stone that He will not be able to raise? Obviously, the attempt to ascribe any descriptor, even extremely expressed, to the Absolute inevitably restricts Him *qualitatively*. We try to «push in» the unlimited semiotic definition into the limited one, *i.e.* to consider it as an object with limited properties. There are also problems in the G. Cantor theory of multitudes, especially in the problem of continuum. As J. Poincare wrote in his «Science and method»: «There is no actual infinity. Kantoreans forgot it and fell into contradiction». Part of problems has been resolved, as they deal with «actual» but not the «potential» infinity, which is meant of as an endless process of per-elementary on-growth of the finite multitude - within the constructive approach in mathematics, where it is prohibited to «put the infinity in brackets». However, the «time» properly put in brackets as a condition of the possibility for the *endless process*, and it gives rise to new contradictions. «Carring out constructive processes, we often come accross the obstacles connected with a shortage of time, place (space) and material. Nevertheless, further, we will not consider them in our speculations about constructive objects. We will think in a way as if they would not exist, and as if space, time and material - required for the realisation of next step in the process under consideration - would be at our disposal. Acting this way, we will digress from the limitation of our possibilities in space, time and material. This digression is commonly named as *abstraction of potential realisation*» [11]. But, if the «potential realisation» is refused in rendering the status of reality (evading this wayparadoxes of «actual infinity»), then quantum mechanics do not deal with objective reality anyway, as it describes only «abstract realisation» of the possible reality manifestation under measurement.

Probably, one should agree to the thing that, semiotically, it is correct to define objects in their relation to each other (interrelation), but not global categories which are not proper signs. An attempt to describe the subject «descriptorily», - it is like when we describe an organism, individual, individuality, personality, *etc.*, - immediately turns it to the object. Then the opposition «subject-object» disappears and the notion of object becomes indefinite. If there is no opposition, then the object (object reality) is anything without exclusions, - it is Absolute to which there is nothing to oppose. It is due to this, that Absolute cannot participate in any process of semiosis and any of our considerations based on signs and meanings. Analogously, absolutisation of Subject isolates it from Reality and makes it to be God excluding it from any scientific investigation.

Subject - wholistic (systemic) - is the opposition to object (analytical) reality description. The subject, as a total unity, can even not be used in the plural as it is not determined in the semantic space of properties and, hence, we cannot differentiate «subjects» in their expression degree. The subject cannot be analysed with analytical methods of the object approach, *i.e.* regarding «object Being», it *is* Nothing [13]. However, the very quantor of existence, essentially, is *the evidence* about the thing that this object is represented in mind (available for perception and realisation (reasoning)), because somebody has to *testify* it and bring out the real verdict. The systemic analysis, which will be focused on later, is currently the only scientific trend that tries to carry out the wholistic approach to the investigation of reality.

Due to the thing that any scientific method, in modern understanding, is based on this or that system of signs - *i.e.* it is semiotic - then, as it was mentioned above, *limits of semiotics are limits of any scientific method*. It seems that nobody argues about the thing that any method has its limitations (conditions and field of application) as this method comprises a system of rules that determines the conditions of its application and limitations. Logic of a method may be immaculate, but its axiomatics, on which it is based upon, is not the only one.

PROBLEM OF «SUBJECT» IN PHYSICS

Subject in physical paradigm, problem of "observer" in relativistic and quantum physics, problem of wave function reduction.

Very often, *object* description is believed to be the synonym of *the objective* one. It is connected with the thing that objectivity is derived out of «on-the-other-side» independent from a subject's existence, regarding

---

[13] Obviously, it is due to this that it is so insistently excluded from the classical scientific paradigm.

our «subjective» feelings of some Platonean «true» transcedental «space of objects» from which we receive signals, in some unknown way, as feelings. Similarity of our mental «reflections» is, in essence, the ground of their objectivity and alleged independence from the subject. Here, they forget the way in which a fact of nervous disturbance becomes feeling, gives rise to the so-called psychophysical problem which has not still been reasonably solved. But «on-the-other-side» world, it is not neccesary to ground such a similarity at all. General laws of making our mental maps are quite sufficient here. Anyway, acting in accordance with the Okkam maxima: «does not involve excessive entities», which do not let us base, in our analysis, on the «objective world on-the-other-side» of feelings untill it is cardinally demanded by logic.

Although, natural sciences try to exclude the Subject from their objective «deterministic» paradigm, using all possible ways, however, despite all efforts, implicity is present in them. Its first explicit presence, as it was mentioned above, was found in the special theory of relativity masked as «observer», without whom it was impossible to determine the physical countdown system. There, he behaved in a particular way: just as «sacred spirit», he could immediately transform from one system to the other. His world perception (space and time) cardinally changed in each countdown system on quite «objective laws». But if these laws are objective, then the «real objective world» should change depending on a subject's «countdown system», which he should not depend on from the definition. To acknowledge the fact that objectivity depends not only on the «object world» but also on the subject, was impossible, as it had a shade of vivid misticism.

The second relevant phenomenon of the subject occurred in quantum mechanics. V. Heinsberg reported on that: «the concept of «event» should be limited by observation». This conclusuion is very relevant as, obviously, it shows that observation plays a decisive role in an atomic event and that Reality is distinguished, depending on our observation or non-obseravation» [4]. According to Heinsberg [4], «quantum hop» that occurs during observation refers to a change of our knowledge. However, the Copenhagen interpretation keeps suggesting the thing that Schrödinger equation describes something which does not depend on either our consciousness or an observer's existence. But «quantum hop», which happens during observation, from the «possibility to reality» is impossible without the observer. in Hesenberg's viewpoint. Everett notes the thing that it leads to the logical contradiction of Copenhagen interpretation because, on one hand, it cannot do without an observer who cannot be described with quantum formalism. On the other hand, he proclaims quantum description to be completed.

John von Neumann [12] also marked the duality in interpretation of the measurement process in quantum mechanics:

> ...it is absolutely true *per se* that measurement, or the process of objective perception is a new entity regarding the external physical world connected with it and not confined to it...such a process withdraws us from the external physical world or, to be more precise, puts us into the uncontrolled mental internal life of an individual. However, in spite of this, there is a requirement, basic for the whole scientific outlook - the so-called principle of psychophysical parallelism, according to which it should be possible to describe... the «out-of-the-physical» process of subjective perception so as if it would proceed in the physical world, - *i.e.* to compare it to successive stages of physical processes in the objective external world, in the usual space... However, in any case, no matter how far we would go in our calculations - to the mercury vessel of thermometer, to its scale, to the retina or to brain cells - at some moment, we always have to divide the world into two parts; the observed system and the observer. In the first one, we can, at least, principally, investigate all physical processes in any detail; it is senseless in the last one. Quantum mechanics describes just those events which develop in the observed part of the world for the time till its interaction with the non-observed part using process (B); however, as soon as such an interaction occurs, *i.e.* a measurement is realised, it prescribes the use of process (A). Duality is justified this very way. Herein, there is a danger of breaking the principle of physical parallelism, if only we do not show that (understood in the above-mentioned sense) the borderline between the observed system and the observer may be arbitrarily shifted...

Let us note that, principally, it is possible to evade the troubles of «physical parallelism». For instance, Bertran Russell believed that, as psychic feelings are a source of data also in physics, any scientific knowledge is principally confined to that of psychological (anyway, everything what happens in the world is very closer to the psychological explanation).

Finally, most of the physicists had to connect the procedure of reduction of wave function (pointing out one meaning out of a multitude) with subject's perception act or the act of realisation. However, the subject here was still associated with a *living being*, usually with human. Y. Wigner, a famous physicist, even suggested the theory whose main idea consisted in the thing that all consciousless matters evolutionize on purely mechanistic algorithms (*e.g.* Schrödinger equations). However, when a system's quantum state turns to be linked with the state of some «conscious being», some mysterious physical process, that leads to reduction, steps in. In Wigner's contribution [13], there is an even more categoric assertion: consciousness should not only be included in the theory of measurement, but consciousness can affect the reality. Schrödinger formulated a similar idea with the epilogue that the book «What is life? The Physical Aspect of the Living Cell» ends in.

However, not everything goes smooth with this approach as there are reduction processes without the «living being» (by the way, how could it appear without the reduction process?). For example, vacuum fluctuates, virtual pairs particle-antiparticle occur and then they annihilate again. But, if this process proceeds on the interface of the black hole, then one of particles may get there - the other may irradiate (black hole evapouration process). This way, a real object appears from the virtual state. In this version, it is difficult to connect «living being», «observer» and «subject». However, if to remember the subject as the world unity, the opposition to the object-analytical principle of the world description, then the subject is always immanently «present» in the «object» Being (true, being not the object).

The many-worlds interpretation of quantum mechanics suggested by Everett and developed by Wheeler [14, 15] is the most radical variant for this problem's solution. A closed system, including the measured subsystem, the device and the observer (the whole Universe!) is considered in this approach. According to Everett interpretation, each of the superpositon components describes the whole world, and none of them has any advantage over the other. There are as many worlds as the number of alternative results which a considered measurement (observation or perception) has. In each of these worlds, there is also a measured system, the device and the observer. The system's state, that of the device and the observer's consciousness, in each of these worlds corresponds to only one result of measurement; but results of the measurements are different in different worlds. There is no reason here to discuss this concept as, in it, the Absolute (Universe) begins to multiply together with «Subjects» and «Consciousnesses» most radically. The points of how even two «Absolutes» can «co-exist» and what they differ in refer to theology, divine trinity, *etc.*

The analysis of idiological positions, on which base both quantum mechanics and its paradoxes occurred, was presented above. Let us pass on to the consideration of the experimental material considering the formulated viewpoint about the necessity to introduce the observer in the scientific paradigm with all one's perculiar features of perception.

EXPERIMENTAL BASES OF QUANTUM MECHANICS, TYPICAL EXPERIMENTS

Experimental bases of physical model reconsideration regarding Reality, micro-objects interference, micro-objects scattering, rediation transmission through polarizers.

To pass on the formulation of the material for quantum computer, we need to consider the bases of quantum mechanics. A centennial jubilee of this theory was marked not so long ago, *i.e.* in 2000, but its paradoxicality or, sometimes they say, «counter-intuitiveness» keeps on worrying experts a hundred years later after its origin.

Let us consider experimental facts which lay ground for quantum mechanics. First of all, we should mention the result obtained by Millikan who determined the value of minimal electric charge. The particle-carrier of this charge was rendered the name of «electron».

The second proof of the thing that substance structure is discrete was Rutherford's experiments with *X*-rays, which showed that crystalline bodies are atoms ordered and symmetrically arranged in space. The picture analogous to that of usual interference was observed within the visible light in the radiation dissipation with the wavelength corresponding to spacings among them.

The third base for reconsidering the classical outlook in physics was the circumstance that a number of experimental facts could be described using the statement about the thing that radiation is a flow of discrete particles. Their energy turned proportional to new constant Planck-called *h* constant. Later, the Louis de Broglie hypothesis about objects wave functions with the example of electron was confirmed. The certainty of such classical term as «wave» and «particle» was swayed.

And, finally, it was not possible to solve the problem of atom's stability within classical physics thinking that electrons orbit the positive nucleus as the planets around the Sun. One can imagine a number of star systems that orbit the central luminary along different orbits, but each of such systems will be unique; at the same time, any atom of certain element and any part of this atom (electrons, nucleus) have *absolutely identical* properties [14] on the commonly accepted current outlook.

It follows from the above-said that the outlook on the possibility of outer reality description in such terms as *continuity, infinity and differentiability* became contradictious to facts. It was the most serious crisis in physics of the XXth century. Something had to be done, first of all, with our outlook of the outer world. It is impossible to understand the outer world within old viewpoints if we recognise the thing that experimental facts imply reality. It is necessary to note that there are problems connected with it not only in physics, but also in mathematics using these terms. The presence of common unresolved paradoxes allows us to think that there are also common reasons for that.

Let us consider the experimental results that most brightly express the peculiarities of quantum phenomena. Suppose, to begin with a well-known experiment mentioned in, practically, every book [16, 17] - occurrence of interference image in the radiation dissipation on two slits (Fig. **3.1**). We have a source of, *e.g.*, electrons designated as *I* in the figure. In the experiments, carried out by V.A. Fabrikant and co-authors (in the 40's of the XX century), which lasted for several months [17], a slightly heated cathode that made one-electron flow was used as a source. Particles interaction was excluded in this case; each particle was before slits 1 and 2. Earlier, the photoplate was used as detector, but one can think that we have high density detectors matrix which reflects particles spatial distribution behind the barrier. The signal from this matrix's elements is shown as histogram *P(x)* along co-ordinate *x*.

As a result of scattering of a big number of particles, there will be a complicated distribution *P(x)* shown in the figure on the left *(a)*. If we close any hole in turn, then the distributions will be $P_1(x)$ and $P_2(x)$, as it is shown in the figure on the right, the sum of these two histograms designated as $P_3(x)$ being not at all similar with *P(x)* when both holes are open. These histograms reflect an electron probabilty of being in some point (dot) with co-ordinate *x*. And it means that addition of probabilities, as it takes place in two incompatible events (*e.g.* «head» or «tail» when tossing a coin) does not happen. In two simultaneously open holes, the distribution *P(x)* is not the sum of distributions $P_1(x)$ and $P_2(x)$. One can conclude that, in this case, the electron does not behave like a coin when only one version of the event is realised with some fixed probability. Such behaviour became the base of thinking that the electron has its wave functions here. We have to conclude that the electron is not only a particle but, sometimes, also a wave.

---

[14]In general, when measuring some property of an object, there are different values in each experiment due to experimental precision; however, the mean value - as a point of countdown - is chosen as quantitative for the character under study.

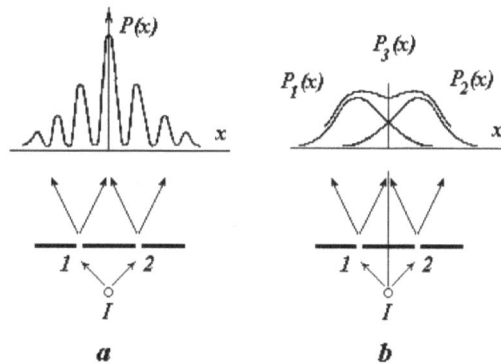

Figure 3.1. The double slit experiment showing electron interference.

As the second typical experiment, let us describe the results of experiments with α-particles scattering (nuclei of atom ^{4}He) and nuclei of atom ^{3}He (Fig. **3.2**). In the figure there are sources of particles I_1 and I_2 and detectors D_1 and D_2. Simultaneous registration of particles by both counters are thought to be the act of scattering. Suppose these are single particles. Let us consider several variants of this experiment.

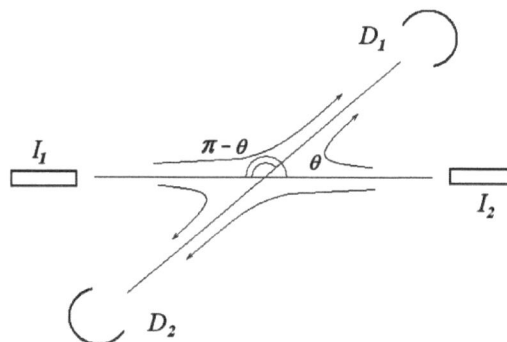

Figure 3.2. Experiments with scattering.

Variant I. One source, *e.g.* I_1, emits α-particles, and I_2 - nuclei of ^{3}He, this element's isotop. Detectors D_1 register only α-particles, and D_2 - nuclei of ^{3}He. Let the events measured probabilty became p in this case.

Variant II. The sources are the same, but now each of the detectors can register both α-particles and nuclei of ^{3}He. In this case the probability of scattering is equal to $(p+p')$ or $2p$ for the angle θ equal to $\pi/2$. It is quite obvious, as the increase of an event's probability is considered by the consideration of scattering act of each of particles into both detectors.

Variant III. Both sources emit α-particles. Herein, in the experiment, the event's probability turns to be equal to $4p$ for angle $\pi/2$, and this is surprising! The probability becomes four times higher, though, if the source's intensity does not change, then it seems that the number of possible outcomes of events - either simultaneous functioning of detectors when a particle contacts it or not, compared to the previous case, would not have to be changed. Thus, in this case, the classical decrease of probabilities does not happen. Analogously, we have to believe that, in variant III, the α-particle is already not a particle but a wave!

Now let us consider one more element with light ([18] and «Lectures on Quantum Computing» by Dan C. Marinescu and Gabriela M. Marinescu, 2003). The following simple experiments can be realised using minimum devices: any strong light source and three polarizers. The experiment demonstrates the bases of quantum mechanics with the use of photons polarization.

We have a light source (Fig. **3.3**) and a detector which measures its intensity on the screen. First, we set the first filter which gives way to only vertically polarized light, the intensity of passed radiation being twice lower. If we set the second filter, which gives way to polarization, perpendicularly to the first one, then light intensity will be zero behind it.

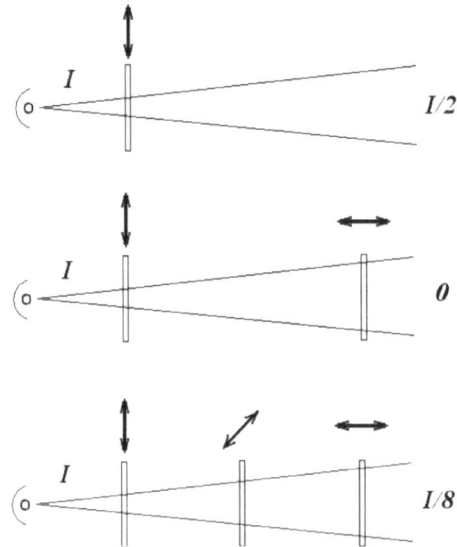

Figure 3.3. Experiment with photons polarization.

One can hypothesize that the filter functions so that it can «screw up in» randomly polarized radiation to one of mutually perpendicular axes. Then the output intensity will be equal to one and half of the falling light intensity in the first experiment, and the passed photons are now vertically polarized, those unpassed being reflected. It is clear that, none of the vertically polarized photons can pass through the horizontal filter in the second case. Now let us insert one more filter, which gives way to radiation polarized at the angle of 45 degrees, between these filters. It seems that any changes are not to occur in the passed light intensity. The experiment shows that, instead of the dark screen, we will see that a slightly lit intensity of radiation which transmitted through all filters will be equal to 1/8 of the initial. Common sense intuitively implies that adding the filter can only decrease the passed light intensity. How can it increase it?

To explain experimental results, physicists had to introduce such a concept as «wave-particle» and also to establish the principle of superposition. The first one meant that one and the same term (electron, photon, etc) was used to explain principally different situations. They were different in classical and non-classical probability distribution when registrating a measured property. However, it was supposed to be one and the same object observed in different conditions. Taking to the principle of superposition was the next step in making up the explanatory paradigm of quantum mechanics. According to this principle, if observation of some quantum system's states is possible as an experimental result, then a system may be in each of these states simultaneously before observation. It turns out that one and the same object may have different meanings of one property simultaneously, be in different places, *i.e.* have different values of the co-ordinate. Such a radical reconsideration of our classical world outlook was conditioned by the necessity to explain experimental results analogous to the experiment with radiation described above. It is impossible to imagine it, and it is for this reason that quantum mechanics is often said to be counter-intuitive (*i.e.* the primary object reality description algorithm is not adequate to Reality proper).

- *To prove this idea, one can give a long enumeration of quantum phenomena. Their most interesting and complete statement is presented in the book [19]. To those who are well aware of the above-mentioned examples, we propose a brief outline of the current state of things. Interference [19] was registered in numerous experiments with both particles, atoms and their molecules including up to 108 atoms. It expresses itself in the correspondence of quantum system's typical energy, to which a certain de Broglie wave is compared, and spatial dimensions during its observation. The system begins to «sound» only in a certain resonator. The phenomenon proper was well studied but its gist is not clear.*

The principle of superposition accepted for explanation contradicts to our typical world outlook in which an object is always localised in space and exists in temporal measurement. Indicative are experiments with a delayed choice where the photon first fell Mach-Zehnder interferometer, and then the conditions of its expression as a particle or a wave [19] changed. In some space point x_2, the beam divider switched on at the output of interferometer at the moment of time t_2 after the transmission of input divider in point x_1 at t_1. Interference was registered with the presence of a divider at the output which meant transmission of the photon on both ways of the inteferometer. When there was no divider, the photon was registered as particle, and it meant the choice of one way. It is not clear how the photon could «know» about the future random measurement version. It is possible to conclude that our object way of description, unbreakably connected with the spatial-temporal outlook, leads to unresolved paradoxes. The considered examples are vividly indicative of the impossibility of using it to explain quantum phenomena. But it is difficult to refuse our accumulated «baggage».

One has to admit that the objective way of world outlook (vision), being natural to us, is not the only possible one. In this connection, let us try to consider experimental facts from new standpoints which were formulated in the previous paragraphs. The main idea consists in the thing that the world's unity and indivisibility should be described in terms of phenomena. In this case the whole cannot change gradually or in parts. And, if the world is like this, we have to use the corresponding description, though we are used to consider phenomena as evolution of objects in space-time. In this description, conditionally, we have to divide the whole into parts, to indicate them as objects and localise them in the mental map space rendering them a certain set of properties. Alternative to this is the wholistic world description as phenomena (systemic description which has a typical spectral image). These ways of description are completely different and, unfortunately, the same terms are often used in them. For instance, a monochromatic wave with frequency v is endlessly extended in the spatial-temporal vision. In the spectral vision, it has one harmonic that corresponds to it, but there is neither space nor time in this description. One can connect it with the finite energy and it is not clear which energy the infinite wave has. The notion of spatial interval can be used as an interface condition. However, there is no co-ordinate here as a continuous parameter. Here it is possible to speak about the QUASI-impulse which is determined by these borders, but the attempts to use the concept for the quantity of movement, *i.e.* the impulse equal to the product of mass to speed, lead to paradoxes just due to the absence of the movement proper in this description. Everything is alright if we strictly stick to the range of chosen description, but some confusion in terminology obfuscates us.

- *The «object» way of describing the objective has developed during social evolution and due to the need for communication. This way of the world partial division is necessary for its informational presentation to ourselves and other «reference systems», for the comparance and specification of our knowledge (mental maps, as psychologists would say). Earlier, such division of the whole and the indivisible led to paradoxes, e.g. those of Zenon; it turned to be just impossible at the quantum level and numerous experiments are indicative of this thing.*

  Various illusions demonstrate the thing that the perception result may be mixed, i.e. there is always a probability of error in pointing out an object. Based on this, it is possible to presuppose the existence of certain regularities, e.g. visual perception [20, 21], Fig. (3.4).

Figure 3.4. Distinguish the object from the background.

In the picture, there is a black-and-white film without semi-tones, and the task is to recognise (distinguish) the object. Classical approach implies the following algorithm. Our «board computer» gets to work that proceeds in about this way: first an image is scanned, pre-grouping of spots on the size, location, etc, the supposition about the possible type of an object is extracted from the «datasbase», then the chosen group is delineated (drawing a borderline - a conditioned line; operations of interpolation, differentiation are carried out to distinguish the object from the background), then we compare the result to the supposition followed by possible correction, estimate of coincidence, acceptance of and abandoning the hypothesis. It is clear that the computing process, the way it is presented here, is the function of classical computer. Completion of counting ends in a change of an observer's psychic state, but, it is already beyond the «authorities» of this «computing machine».

In principle, an observer may not realise the above-described process and not know such terms, not even mentioning the thing that the majority does not think that there are certain perception algorithms. Recognition of an object proper is not always possible if, e.g., there is no image of «Dalmatian in the moonlight» in the «database».

As an example, distinctions between them, i.e. between what is in reality and what we imagine in its perception, will be demonstrated by successive pictures of the faraon's rotating golden mask.

$$(gif - animation \ll http : / / ctpax - cheater.narod.$$
$$ru / htmldocs / pefs / pm3d2.gif \gg)$$

There is no «concave» face in our database; so, the internal part of the mask looks concave anyway. A change of the mask's rotation direction and distribution of its illumination, when it is directed to the observer with its internal side, occurs for the accordance of perception.

Figure 3.5. Is not really surprising why cannot we see the concave face?

We can find hundreds of examples for optical illusions the perceptive paradox of which is indicative of the thing that the existing settings of receiving information and signal processing contradictious to experimental facts.

How will our description change if we abandon the object representation?

Let us, once more, consider experimental results that express the peculiarities of quantum phenomena.

Let us return to the experiment with two slits (Fig. (**3.1**)). If you close any of them in turn, then there will be distributions $P_1(x)$ and $P_2(x)$, as is shown in the right figure (*b*), the sum of these two histograms, designated as $P_3(x)$, is completely unlike $P(x)$ when both slits are open. One can conclude that, in this case, it is impossible to present the electron as an object (particle). The term «wave» is best suitable here to describe experimental results. True, it is unclear «which» wave and «what on». But, in the case of only one slit, electrons behave as a flow of particles, *i.e.* objects localised in space, having their trajectories, and that can be temporarily described. It turns out that wholeness of the world is manifested in such a phenomenon as interference, and our attempts fail to characterise it using the common object description. That is why there is a replacement of notions in the statement that, in this variant, the electron has wave properties.

The second typical experiment - in its third variant - with the scattering of nuclei of atom ^{4}He (*i.e.* α - particles; here the term «particle» has become «tightly fixed» irregardless of their behaviour) and nuclei of atom ^{3}He (Fig. (**3.2**) - is confined to the previous case of interference, if to believe that the role of slits 1 and 2 is performed by sources and particles proper. The situation is classical if it is possible to outline the trajectories and, hence, to pass on to the object description. In other cases, objective description is not functional, it is not our choice. We can only change the observation conditions which make us transfer from one representation to the other.

Let us turn to the consideration of experiments with radiation and polarizers. They allow us to conclude that such property of photons as polarization (or spin) is reasonable regarding a concrete polarizer, but not in general; *i.e.* the property may be considered in some basis and can be measured by us related to it. It turns out again that wholeness has no traits, and it is possible only in experiments (roughly speaking in «objectivization» of the phenomenon) that a certain quality can be outpointed and its quantitative characteristics can be obtained. Figuratively, the photon has no «documents»; *we just ascribe* these properties quantitatively expressed by observation results of the whole phenomenon (situation).

- *The proposed description allows us to exclude the duality of «wave-particle» and to get rid of the superposition principle when considering quantum phenomena. Logic may be restored, but the price for this is the necessity of reconsidering our own world outlook. The mathematical apparatus of quantum mechanics, surely, remains the same, but its interpretation changes.*

SEMIOTIC ASPECTS OF PHYSICAL PRINCIPLES

Conditionality of object-based description of reality, polysemanticity of sign and role of context, multiplicity of interpretations of linguistic description, probability of meaning actualization, problem of context-independent description of object, "superimposition" principle in psychology

Werner von Heisenberg, an eminent physicist, once noticed [4]:

... understanding of any kind, being scientific or not, depends on the thing that we can transmit our thoughts. Any description of events and their results are also based on language as the only means of understanding. Words of this language express notions of everyday life, which can be (words) specified in the scientific language of physics to the notions of classical physics... As soon as a physicist abandons this base, he would lose the possibility to formulate his ideas uniquely and will not develop his science on.

It seems totally obvious that any knowledge is translated in some semiotic system. Sign systems are the subject of semiotic science and, hence, semiotic «laws» may be implicitly involved in those laws formulated within the science using it and considerably suggest itself in our reality outlook. Therefore, we have to focus in more detail on the way of object semiotic reality outlook and its limitations.

As it was mentioned above, accepting the «signal» origin of our senses aimed at the orientation in Reality [15], being on the other side of senses, we also have to acknowledge *the model, semiotic* character of our outlook on it. It is obvious that, using signs, it is possible to make up a mental map of the thing that we call «external surrounding», but the map is no territory but the object, it is not the thing «on-the-other-side» of senses, but just only a semiotic representation of some stable complex of senses which has meaning for us. The point about the thing if some «thing» on the other side of senses corresponds to the object presented as a sign (or signs) on the mental map is excessive. Is Reality broken into separate «self-sufficient» bits, or is it one of the ways of its representation and ordering of our senses, convenient only to a certain limit? We cannot enter this «on-the-other-side» world and, thence, hasty conclusions about it may lead us to a logical deadlock. Postulating the existence of some «absolute thing» (in the Platonean sense) and its semiotic description as an object, as a relative and not precise representation, we implicitly introduce extra-limitations, having no sufficient bias in our theories. Okkam used to say: «it is not worth producing entities without necessity».

Describing an object is usually realised by enumerating its qualities and their intensities: *e.g.* the electron has its mass, charge, spin, speed, *etc.* Some characteristics are unique (*e.g.* charge), others are not (*e.g.* impulse). At the first stage of determining the mental map's model equivalent, the most obvious and natural is *mental space reflection of the «first signal system» and the vector space of properties of the «second signal system»* and, essentially, it is realised by linguists in the situation of direct [16] (absolute) definition of objects through their properties: $\Omega \rightarrow \vec{U} = U(Q_1, Q_2, ... Q_n)$. Herein, object Ω is reflected to vector $\vec{U}$ in the space of properties $\{Q_1, Q_2, ... Q_n\}$. It is obvious that object description cannot be univocal (monosemantic) because meanings of properties are not strictly definite. For instance, an electron may have its spin of both +1/2 and -1/2. Semiotic polysemy is well known in linguistics: *e.g.* in English the word «table» may have the meaning of directly «table» (piece of furniture) and «chart», «boot» may be used in the meaning of footware and «loose cover», «pit», «trunk», «profit», *etc.* Out of the context, a sign's meaning is not completely determined, and semiotic description of an object out of a certain situation is «virtual», as it has some multitude of permittable «states» or meanings. The analogous phenomenon is also typical of the first signal system. Let us consider some examples of visual perception.

In Fig. (**3.6**) (*a*) you can see the book opened in the direction to or from us. In Fig. (**3.7**) (*b*) it is possible to see either a ladder or an overhanging cornice and, in Fig (**3.6**) (*c*) - either an old or a young woman.

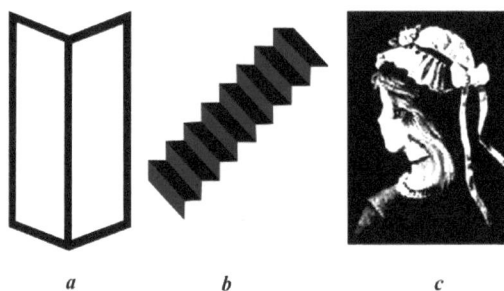

a   b   c

Figure 3.6. Mach's figure (**a**), Schroder's ladder (**b**), Boring's Young Girl/Mother-in-law (**c**).

---

[15]At this stage, we would not like to discuss what is understood by the term «objective reality». Let us confine it to our intuitive view that it is a source of our senses independent from us and, further on, we will indicate it as «Reality» and write it with the capital letter.

[16]Indirect (relative) definition of an object is carried out when compared to other objects using such tropes as metonymy and metaphor. Absolute definition is presented by enumeration of an object's properties and their intensities (usually in dictionaries).

Particularly, the sign proper can be perceived in a different way. For instance, the sign in the centre, Fig. (**3.7**) (*a*), can be read as both a letter and a number depending on the vertical or horizontal read-out. In Fig. (**3.7**) (Escher), there is an M.C. Escher's engraving on which we may perceive either angels or demons depending on what we refer to the background or image.

Thus, *before the act of perception*, an object's description will be multiple and it is rather «virtual» than «real». At probability p_1, it may be perceived in the state (or meaning) Ω_1 and in the state Ω_2 - at probability p_2. Heinsberg discussed such a quantum state interpretation: «If, on the other hand, to understand the word «state» in the sense that it denotes rather a possibility than reality - it is even possible to replace the word «state» by that of «possibility»; then the notion of «co-existing possibilities» seems quite acceptible, as any possibility may include other possibility or overlap (cross) with other possibilities» [4].

Let us note that different states may, in principle, «interfere» with each other: *e.g.* in Fig. (**3.6**) (*a*) it is possible to see not only a «volumous» book opened in the direction to us, but a flat figure shaped as an arrow; and Fig. (**3.6**) (*b*) may be perceived not only as a ladder or a cornice, but also as a paper ribbon folded like an accordion. In the gif-animation (Right and Left.gif: ≪ *http : / / aralbalkan.com / 1058* ≫) you can see a «volumous» ballerina that spins either clockwise or counter-clockwise. With a more carefull look, you can also see a small figure that makes fluctuating movements.

Multitude of object interpretations is typical of not only physics. Sociologists, when polling to know public opinion, divide the investigated population into social groups thinking that the similarity in lifestyle and burning needs leads to unification of perceptions and interpretations of information, also the attitude to it. However, besides social factors, there are also psychological, personal, individual-typological, cultural and a number of others. An attempt to make them formally average may lead to just absurd results. Let us consider a simple example, the image in Fig. (**3.7**) (*b*). Depending on the way an individual interpreted it (like a pyramid or a rosette), he or she may perceive either 6 or 7 cubes. Formal evaluation of the averge opinion leads to the paradoxical result - 6.5 cubes, which does not correspond to any real perception! Hence, if we want to picture (imagine) an object *per se*, *before the act of perception*, we should describe it as *a multitude of potential meanings (values)* without averaging. Naturally, possible interpretations are not equi-probable and considerably depend on the context. Obviously, an object's description by different individuals may be different from each other to a certain extent, and we need a formal procedure which could unify various interpretations focusing on the most considerable general moments which we could agree to. Possible variants of semiotic meaning are properly defined by this.

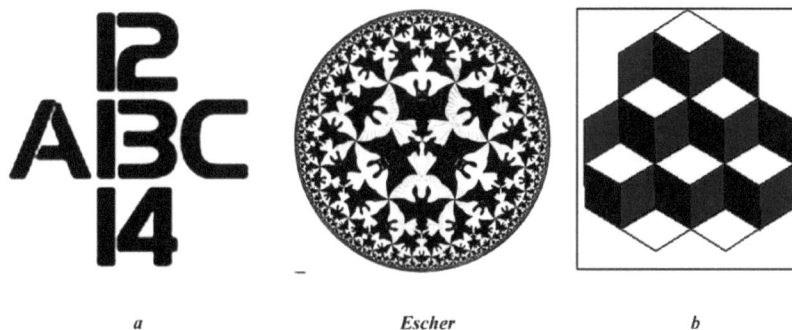

a   *Escher*   *b*

Figure 3.7. Letter or number (a), M.C. Escher - <Circle Limit IV (Heaven and Hell)>, 6 or 7 cubes (b).

Obviously, within one mentality, *i.e.* in individuals living in similar conditions and having similar needs, a general meaningful semiotic communication system is to develop [17]. Besides, all notional interpretations of

---

[17]Signs have semantics only when they embody set meanings stuck to them.

objects should be realised with a *certain (definite) probability* [18] within homogeneous mentality. It is connected with the thing that actualisation of this or other interpretation depends on the context of concrete situations whose frequency is determined by common living conditions and surroundings of these people. If these conditions are diferent, then the adequacy of communications is broken (people's understanding each other). As an example, we can remember the behaviour of Eliza Doolittle, a semi-literate (low-educated) flower-girl being in the high society of doctor Higgins after «Pigmalion» by Bernard Shaw. Multiplicity of interpreting any notion is well known to linguists [22] and partially presented in explanatory dictionaries. Usually, fields of the most probable usage of certain meaning are indicated in them (*e.g.* common colloquial, dialect, science, *etc.*). The attempt to describe an object as «self-sufficient», *independent from conditions of a subject's perception*, inevitably makes it polysemantic and «virtual».

Psychosemantic analysis allows us to approach to the points of defining the variants of notional interpretation of perception in a more precise way. As each object (any phenomenon perceived by a subject is understood as «object») can be described through expressivity of properties, it can be represented by vector $\vec{U}$ in the space of properties with coordinates $\{Q_1, Q_2, ... Q_n\}$, where $\{Q_i\}$ - expressivity (intensity) of *i*-property or as a «profile of properties» shown in Fig. (**3.8**).

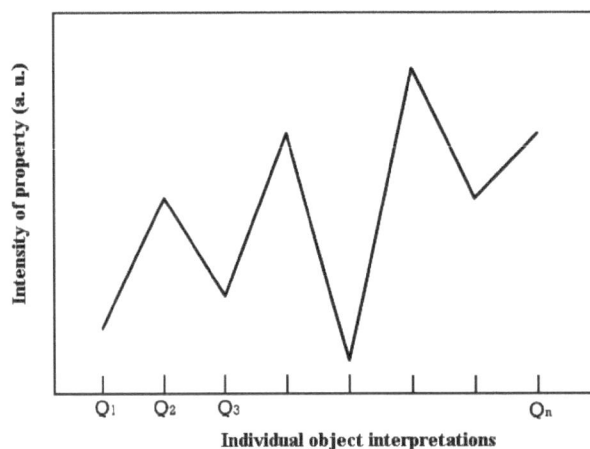

Figure 3.8. Object profile representation.

Suppose we have *N* of individual object interpretations: $\vec{U}$ (*m* = 1, 2,..., *N*) for each of *N* randomly chosen representatives of some mentality. We have to point out the variants of most consensual interpretations and to estimate the probability of their actualisation in perceiving an object by a random member of this group. Let us choose the coefficient of correlation among their profile-descriptions as a measurement of similarity. In this connection, we can norm vectors $\vec{U}_m$ without any limitation of their commonality and think them single ($\vec{u}_m = \vec{U}_m / |\vec{U}_m|$). As it is simpler to analyse semantically independent interpretations, the method of main components [19] is the most convenient to be used to point them out [23]. For this purpose, let us calculate correlation matrix (**R**) for all interpretations and solve the characteristic equation: $\hat{\mathbf{R}}\vec{s} = \lambda\vec{s}$. Herein, $\hat{\mathbf{R}}$ is matrix operator, λ - own value of matrix-operator ($\hat{\mathbf{R}}$), $\vec{s}$ - own vector (main component). As the main components $\vec{s}_1, \vec{s}_2, ..., \vec{s}_k$ are orthogonal factors connecting the bundles of most correlated descriptions (see Fig. (**3.9**)), they can be a generalisation of possible semantic interpretations $\vec{u}_m$ (*m*=1, 2,..., *N*).

---

[18]It is also important for adequate understanding of individual behaviour by other «subjects». If an individual uses unlikely interpretations, he or she looks, at least, strange. *E.g.* a schizophrenic unites such objects as «pencil» and «boot» out of three: «pencil, paper, boot». Formally, it is logical. However, such interpretation is usually unlikely by mentally healthy people and, thus, most often: it remains implicit (latent).

[19]See **Appendix B**

Each vector $\vec{s}_m$ sets some object (state) value, common for the whole mentality (its permissible interpretation, common for all individuals). Thus, we replace N of an object's individual evaluations by K of *consensual independent interpretations* of this object ($K \le N$). As own values of λ_i determine the contribution to dispersion of the corresponding component, *i.e.* the «power» of vector descriptions bundle connected with this factor [23], and $\sum_{m=1}^{N} \lambda_m = N$, then value $P_m = \lambda_m / N$, actually, estimates the probability of interpretation $\vec{s}_m$. There is no necessity to consider low probabilty states that reflect particularly individual aspects of values; therefore, the number of considered own vectors K is usually much less than that of the tested N.

Thus, *before the presentation of stimulus* to a concrete respondent from this very mentality, his or her opinion, regarding some object, can be presented as a peculiar «superposition» of possible orthogonal semantic states $\vec{s}_m$ with the corresponding actualisation probabilities - p_m. To illustrate this, let us consider a concrete psychological example.

The attempt to describe a personality in psychology as some stable «profile» of personal features or properties leads to a number of logical contradictions. The thing is that personality as a system, depending on external and internal factors, may be in different psychological conditions, realise various aims and, in this connection, demonstrate different behaviour that does not fit the Procrustean bed of this rough profile. As an example, let us analyse a mid-manager's behaviour. Obviously, his or her work requires such qualities as «subordination» and «imperiousness». Usually, they are percieved as opposite poles of one axis as they are mutually exclusive, *i.e.* they cannot be expressed simultaneously. Normally, test methods determine how often these personal qualities are expressed; their total (sum) results in the expressivity of some tendency. However, the mean value does not take into account the thing that these properties cannot be expressed simultaneously; though their expressivity may be considerable, the mean value may be equal to zero, which is absurd. Note that *the probability* of some personal property's expression is not as much the quality typical of personality *per se*, as it is *the characteristics of his or her surroundings*: in particular, if a manager has to «come on the carpet» to his superiors, expression of «imperiousness» is not suitable here, and the situation is quite the reverse for a top-manager. Due to this, probabilities of these qualities expressivity will be changing during career. It is obvious that both this and other quality are very necessary for adequate social adaptation and may be quite expressed in their intensity. However, their expressivity *potential* is, surely, unequal and strongly depends on social conditions.

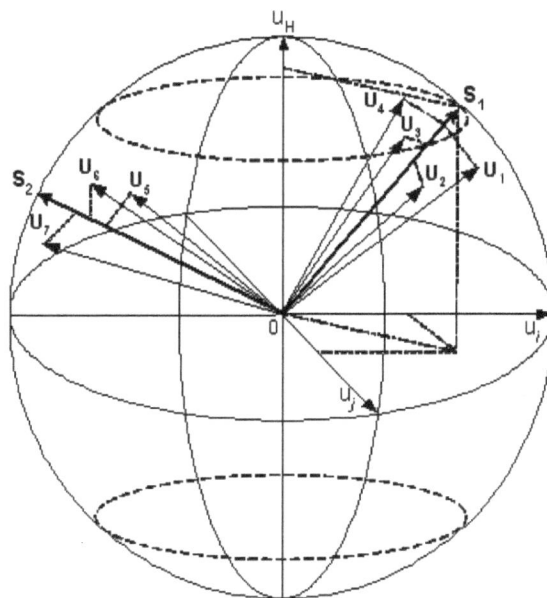

Figure 3.9. Determining the object (S) states on evaluation of (U) results.

Thus, objective description of a personality can be realised only when considering his or her behaviour in *most different conditions*. As physiological and genetic constitutions set one's own spectrum of individual adaptive means, it leads to the development of typical latent personal profiles whose actualisation probability is determined by extrenal conditions of one's existence. Genetic norms of each individual response allow him or her to develop one's own unique psycho-physical and social-psychological means of adaptation towards surroundings, which are most suitable for most relevant and widely spread situations.

- *To check this hypothesis, we used the Timothy Leary test [20] Fourty-five different descriptions of ten testees were obtained on eight standard scales. As the description of each testing is based on one's own experience of communication with a testee in concrete conditions, then, in total, all the descriptions allow us to get the basic individual profiles of personality for each testee in various situations. True, if we make up the distribution of a testee's esimates in the space of two factors: «superiority- inferiority», «aggressivity-friendliness» calculated with standard formulas, then we will obtain not a normally (regularly) distributed two-dimensional random variable, but a polymodal distribution with typical peaks of personality's permissible states. As an example, let us analyse the investigation results of one of the testees (see Fig. (3.10)).*

  Projection of evaluations distribution density onto the plane of personal properties determines the areas of permissible states. In Fig. (3.11) one can see that all areas of most probable personal states (closed lines in darker colour) are additional, as they cannot be realised at one and the same time because mutually opposite properties exclude their simultaneous expression. It is analogous to the principle of complementarity in quantum physics. In the same figure, there is a vector of all evaluations, which, in the best case, indicates only one of the permissible states and, most often, it has no psycological correlation with this personality. Standard vector representation of states with their indicated actualisation probability is shown in Fig. (3.12).

Bivariate Histogram

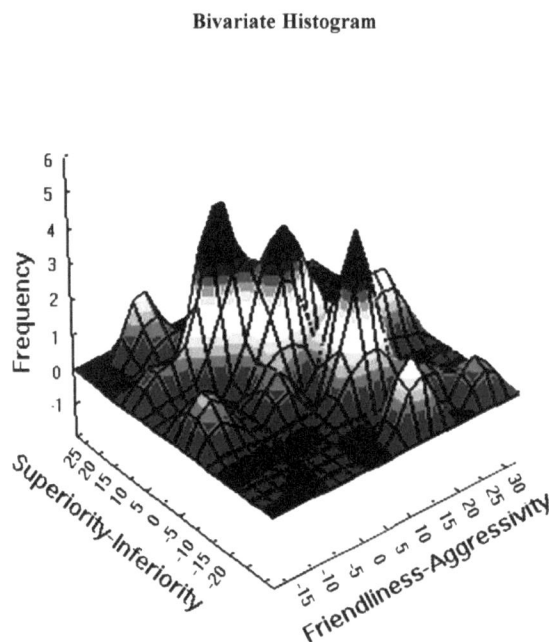

Figure 3.10. Frequencies of individual's personal states of expression.

---

[20]T.Leary, R. Laforge and R. Suczek «The interpersonal diagnosis of personality». A version published by G.S. Vasilchenko (1978) was used.

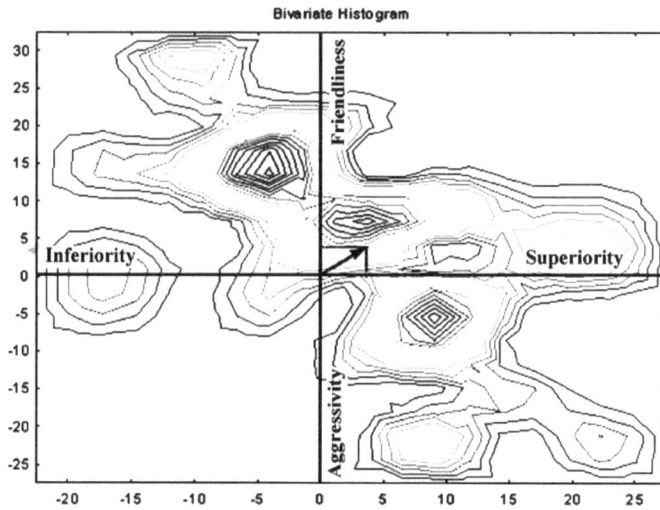

Figure 3.11. Variants of personal behaviour in different states.

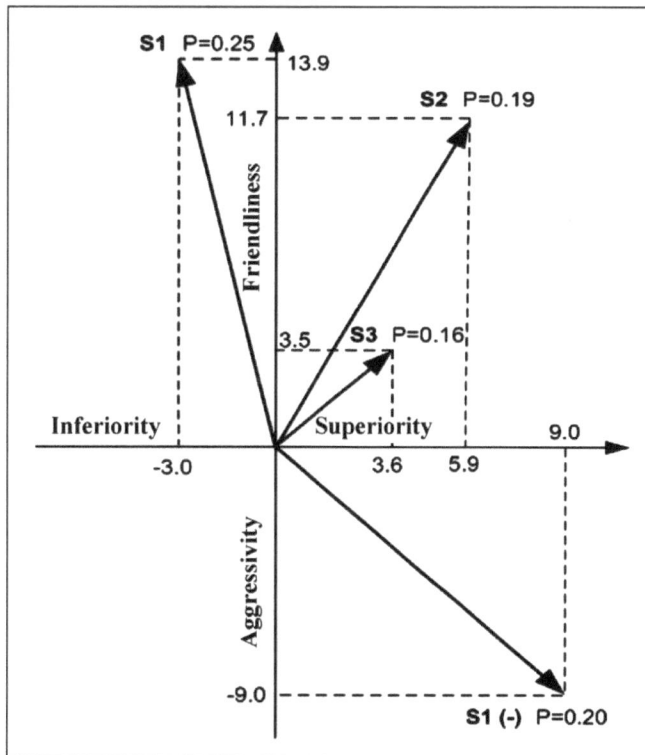

Figure 3.12. Various personal states of individual and probabilities of their actualisation.

Note that each state $\vec{S}_i$ may integrate the bundles of vectors $\vec{u}_i$ having opposite directions (see Fig. 3.13), i.e. the own vector has the positive $\vec{s}_i^{(+)}$ and negative $\vec{s}_i^{(-)}$ component. Such situations upstair, e.g., in sociology when parties or political leaders are estimated by supporters and opponents. In this case, antipodal notional interpretations of objects are often noticed with respondents. Characteristics (description) of any personality may also include mutually exclusive polarities and even be considerably determined by them. For instance, cyclothymia is expressed in just stable fluctuations of mood - from joy to sadness. If the «power» of the back-bundle is high, then it is necessary to consider the correspondence of states separately.

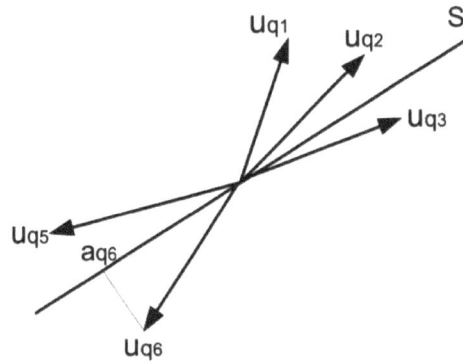

Figure 3.13. Indications: u_{qi} - interpretations of separate respondents; s_i - main component; a_{qi} - factor value (weight).

An attempt to define a complex system as an object out of the perception context inevitably leads to probable description. Description of a personality also becomes «virtual» because object description is invariant in relation to conditions. We have to mention not only possible states characterising personal behaviour, but also how they are expressed in the range of external conditions. Actually, we indicate various permissible variants (profiles) of personal behaviour and their expression frequencies. Obviously, in a real measurement (testing) of personality, we are bound to find it in one of the permissible states. But, *before measurement* and determination of concrete conditions, settings and aims, one's object behaviour will be described as virtual «superposition» of states $\vec{s}_i$ realised at probabilities p_i; each of them may have one of mutually exclusive variants $\vec{S}_i^{(+)}$ or $\vec{S}_i^{(-)}$ realised at probabilities $q_i^{(+)}$ and $q_i^{(-)}$, respectively.

In particular, for example, personal behaviour will be characterised by probability $p^{(-)}$, in which the left pole of the scale «superiority-inferiority» is, and by probability $p^{(+)}$ - the state in which the right pole of the scale («imperiousness») - is realised. In principle, it is analogous to the presentation of electron which, at probability $p^{(+)}$ may be in the state with spin +1/2 and at probability $p^{(-)}$ - with spin -1/2. If, intuitively, we want to preserve the understandable and common object representation on the mental map under polysemy, we also have to preserve its vector reflection in signal system II. However, it will not be a space of properties, but the one of possible states. Not the intensity of properties will be co-ordinates of an object's presentation-vector, but probabilities of these states. Let us try to construct such a space.

Any property is commonly set as some scale U_j. Having chosen some countdown point (0), we conditionally divide it into two poles - positive and negative [21]. Having obtained a number of object's descriptions under its multiple exhibition, we find some multitude of states for it (interpretations): $\vec{s}_1, \vec{s}_2, ..., \vec{s}_k$, realised with corresponding probabilities $p_1, p_2, ..., p_k$. We could define the object as a single vector $\psi(\vec{s}_1, \vec{s}_2, ..., \vec{s}_k)$ in the space of states with coordinates $\sqrt{p_1}, \sqrt{p_2}, ..., \sqrt{p_k}$. It is obvious that the sum of its squared components is equal to 1 (vector's squared length). However, we need to take into account the thing that each state may be realised in two mutually exclusive ways $\vec{s}_i^{(+)} = \vec{s}_i(+\sigma)$ and $\vec{s}_i^{(-)} = \vec{s}_i(-\sigma)$, at probabilities $p_i^{(+)}$ and $p_i^{(-)}$, respectively, *i.e.* the state of the object disintegrates into two components $\vec{s}_i^{\,1}$ and $\vec{s}_i^{\,2}$: $\vec{s}_i = \begin{pmatrix} \vec{s}_i^{\,1} \\ \vec{s}_i^{\,2} \end{pmatrix} = \begin{pmatrix} \vec{s}_i(+\sigma) \\ \vec{s}_i(-\sigma) \end{pmatrix}$

It is possible to decompose the ψ_i the following way:

---

[21] In principle, it is not obligatory, but we will consider this variant as the most frequent one.

$$\psi^2 = \left(\sqrt{p_i^{(+)}} + i\sqrt{p_i^{(-)}}\right)\vec{s}_i \left(\sqrt{p_i^{(+)}} - i\sqrt{p_i^{(-)}}\right)\vec{s}_i^+ = \alpha\vec{s}_i \cdot \alpha^*\vec{s}_i^+, \text{ where } \vec{s}_i^+ \text{ - Hermitian conjugate } \vec{s}_i \text{ and } i^2 = -1.$$

Note that, on Argan plane, $\psi_i{}^1 = \sqrt{q_i^{(+)}}\vec{s}_i$ and $\psi_i{}^2 = \sqrt{q_i^{(+)}}\vec{s}_i$, become not opposite, but orthogonal - like spinors, and it allows us to preserve their vector representation $\psi(\vec{s}_1, \vec{s}_2, ..., \vec{s}_k)$.

We can see that an object's psychosemantic representation, in principle, is similar to quantum-mechanical description, and its vividly not a random coincidence, as ways of semiotic Reality reflection (representation) are universal and they have developed during the long-time biological evolution of genus *Homo sapiens*. W. Heinsberg wrote about it, thinking over logical and linguistic structures in connection with paradoxes of quantum mechanics (interference and reduction of wave function): «These structures may develop, *e.g.*, due to associations between certain intermediate meanings of words; thus, for instance, the secondary meaning of a word, that was almost not left in our mind, may still considerably affect the content of a sentence when this word is pronounced. The fact that any word may cause many half-realised movements in our thinking, may be used to express certain aspects of Reality more clearly using language than it was possible using a logical scheme [4]».

Let us note that *the semiotic system* (language) is not simply a set of signs, but it also implies logical relations among them. Judging by all these things, many meanings of words may be not only experimentally investigated in mentality, but theoretically calculated. Anyway, many terms are developed on the base of comprehension [22] in science, which linguists have long been focused on using it for the profound analysis of notional structures. In the cognition of the world, man opposes various groups of concepts and, pointing out elementary ones, that differentiate semantic units (figures after L. Hjelmslev) and constantly opens new meanings and implications in signs.

- *For instance, let us analyse the following number of words: 1) father; 2) mother; 3) son; 4) daughter; 5) uncle; 6) aunt; 7) nephew; 8) niece. When we oppose words 1, 3, 5 and 7 to words 2, 4, 6 and 8, we will get their distinctive figure: «male - female person». In the opposition of groups 1, 2, 5 and 6 to 3, 4, 7 and 8, we will get the figure «genetically direct - non-direct inheritance». When opposing group 1, 2, 5 and 6 to that of 3, 4, 7 and 8, we will get the figure: «ancestors - descendants». From now on, within the semiotic multitude, the meaning of any element may be presented as a set of three elementary semantic units. So, the word «father» is divided into three figures: «male person», «direct inheritance», «ancestor», and the word «niece» - into: «female person», «non-direct inheritance», «descendant».*

Defining the meaning of a word, we oppose it to new entities again and again and, thus, we open new implications. A word's notion is never complete, as, finally, it gets into understanding the world and ourselves. Word is an inexhaustible source of implications and meanings. As Umberto Eco said in one of his interviews: «We surround ourselves with descriptions of symbols and signs. Everything that exists and what has not been explained yet is sign. Semiotics finds and preserves the commonality of all things and phenomena. There are few things, more interesting than this one, to deal with in the world».

FORMAL LANGUAGE OF QUANTUM MECHANICS

Vector description of a quantum object's state, probability amplitude, rules to deal with probability amplitudes, mathematical description of experiments.

In this paragraph, the formal language used in the description of quantum phenomena is presented. It is a powerful instrument whose advantage is a high level of abstraction from the essence of an analysed

---

[22] Comprehension, «coverage», «extension» - classification and its result, *i.e.* a class of all supported thought objects for which a certain word may be correctly used, from the thing that if these objects exist in reality or not, if their existence is known or unknown.

situation. Due to this quality, it may be used even in this case when there is no precise interpretation of symbols. Let us try and demonstrate this with the example of quantum mechanics.

There is a big number of abstract mathematical theories which, it would seem, have nothing in common with Reality. But, when a new field of knowledge appears and it needs mathematical description, it turns out, as a rule, that, already, there is a suitable formal language most convenient to be used in this particular case. P. Dirac, in his days, introduced the designations and rules of describing quantum phenomena based on linear algebra [24].

- *Let us agree to denote any state of quantum system using vectors «bra» $\langle |$ and «cket» $| \rangle$ (from the English word bracket). The following operations are permissible in the corresponding space formed by these vectors:*

  $\langle x | \times | x \rangle \equiv \langle x \| x \rangle \equiv \langle x | x \rangle$ - scalar multiplication of vectors bra and cket, the result of which is a real number. Multiplication of vectors bra (cket) to complex numbers α , β and additon of vectors bra (cket):

$$\alpha |a\rangle + \beta |a\rangle = (\alpha + \beta)|a\rangle$$
$$\alpha |a\rangle + \beta |b\rangle = |r\rangle$$
$$\langle b | \{\alpha |a\rangle\} = \alpha \langle b | \ a\rangle$$
$$\langle c | (|a\rangle + |b\rangle) = \langle c | \ a\rangle + \langle c | \ b\rangle$$

  Let us agree to indicate the systems transfer from some initial $\ll s \gg$ to the finite $\ll f \gg$ state as $\langle f | s \rangle$. This product of vectors, in general being a complex number, is commonly called «probability amplitude» because the square of moduleis $|\langle f | s \rangle|^2$ is identified with the probability of the present event. Why do we have to use the concept of «probability amplitude», but not «probability», as we did earlier? First of all, it is for the correct description of the results of our experimental data. It is in this that the difference between classical and quantum mechanics is expresed - between phenomenological and processual descriptions.

  Thus, for the case of scattering on two slits (Fig 3.1), it is possible to state the thing that there are two states $\langle 1 | I \rangle$ and $\langle 2 | I \rangle$. Earlier, it was mentioned that there is a postulate in quantum mechanics: if, as an experimental result, it is possible to observe some states of a system, then it may be in each of these states simultaneously. Now the use of the concept «simultaneity» clears out the situation. When considering some phenomenon, introduction of spatial-temporal relations is an attempt to traslate its content into the process. Note that such relations were not needed in the mathematical registration form.

  As an explanation, we can add that the notion of superposition is well known in mathematics; it is used in solutions, e.g., of linear algebra and, in particular, for diffrential equations. If there are two or more solutions $y_1(x), y_2(x)$ for a homogeneous equation with constant coefficients:

$$\frac{d^n y}{dx^n} + a_1 \frac{d^{n-1} y}{dx^{n-1}} + \ldots + a_{n-1} \frac{dy}{dx} + a_n y = 0$$

  then their sum $c_1 y_1(x) + c_2 y_2(x)$, at arbitrary constants c_1 and c_2, is also a solution. As the apparatus of linear algebra and differential equations are used in the description of quantum systems, the term «superposition» is easily included in the paradigm of this theory.

  However, in quantum mechanics, the principle of superposition has the following «physical nuance». The conclusion about superposition of an «object's» states is made on the base of measurements in the same conditions of «object's» ensamble. When there is intensity distribution like P(x) shown in Fig (3.1), then, in this case, it is said «post factum» that each of objects had superposition of states before measurement, i.e. we never can observe a quantum object under superposition. From the principle of comlementarity mentioned by N. Bohr, it just follows that properties of a quantum object may be determined only in the totality of observations. Actually, when analysing superposition, we outpoint the case of distribution P(x) which cannot be obtained in the observation of classical objects. But, if to accept the fact of wholiness of a phenomenon under consideration, there is no necessity for the superposition principle. And, on the contrary, following the principle of superposition and object description, we have to admit that one and the same object may be in two places (points) simultaneously. It can be neither understood nor reasonably explained. Therefore, we have to accept the thing that quantum mechanics is just an instrument of experimental results description.

In the description of quantum phenomenon, let it be set with a corresponding function. The properties of this function are in the content of information about expression probability when measuring the value of characteristics to be searched. It is necessary for further object representation of a phenomenon as a process. Due to the thing that occurrence probability of this or that outcome is the basic measure of observation results description, then, as it is common in the probabilty theory, the sum of outcomes probabilities (total probability) is normed, *i.e.* it is equal to 1. Herein, it is possible to introduce the notion of orthonormed basis on the base of incompatible events - measurement results of a quantum system's state. Remember that, for incompatible events, probabilties are summed, and they are mutiplied for independent events, *e.g.* the result of tossing two coins. By the way, if all possible measurements or observations of a quantum system's states are known, they are said to form a complete orthonormed basis. It is orthogonal because, during measurement, it is possible to observe only one state at each moment of time (in some system of references). We make it normed on our own based on the replication frequency of the given result. It may be put down as follows:

$$\psi = \sum_i \psi_i = \sum_i c_i \, | \, x_i \rangle$$

where $\sum_i | \, c_i \, |^2 = 1$, for vectors the following condition is carried out:

$$\langle x_i \, | \, x_k \rangle = \delta_{ik} = \begin{cases} 0 & for \ i \neq k \\ 1 & for \ i = k \end{cases}$$

Here, Dirac function δ_{ik} is used.

From now on, we will formally introduce the rules of dealing with probability amplitudes [17] and see what will come out of it. Let there be a set of a quantum system's states (using the common language of quantum mechanics: several alternative and uncontrolled ways of transition from the intial *s* to the finite *f* state. However, if a phenomenon is analysed as a whole out of spatial-temporal description, there are no transitions. Because of the absence of suitable terminology, we have to keep to set-expressions, though they distort the point. In this case, probability amplitudes are summed (addition) on formula (3.1); *i* - indicates a system's *i*-state:

$$\langle f \, | \, s \rangle = \sum_i \langle f \, | \, s \rangle_i \tag{3.1}$$

If the analysed phenomenon may be divided into spatial-temporal stages of evolution, then the system transits from $s \rightarrow f$ through some intermediate state *b* and, thus, it is possible to formulate the transition as a successive multplication of amplitudes:

$$\langle f \, | \, s \rangle = \langle f \, | \, b \rangle \langle b \, | \, s \rangle \tag{3.2}$$

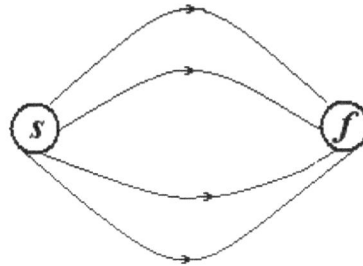

Figure 3.14. Illustration of formula (3.1).

Herewith, the probability of such an event is equal to the product of their probabilities, just as it takes place in a usual case of determining independnet events probability:

$$|\langle f|s\rangle|^2 = |\langle f|b\rangle\langle b|s\rangle|^2 = |\langle f|b\rangle|^2 \times |\langle b|s\rangle|^2$$

The totally analogous way may be used when considering the case of two quantum systems transition - one from state $s \to f$ and, simultaneously, the other - $S \to F$. The resultant probability amplitude will also be equal to the product of amplitudes and, respectively, the event's probability is equal to the product of their probabilities:

$$\langle fF|sS\rangle = \langle f|s\rangle\langle F|S\rangle \Rightarrow |\langle f|s\rangle|^2 \times |\langle F|S\rangle|^2 \tag{3.3}$$

If there are some finite states of a system ($f_1, f_2, ... f_i, ..$) and the probability of its transition to any of them is considered, makes no difference, then, in this case, the resultant transition $\left|\langle f|\ s\rangle\right|^2$ probability is the sum of probability of transitions to various finite states:

$$|\langle f|s\rangle|^2 = \sum_i |\langle f|s\rangle_i|^2 \tag{3.4}$$

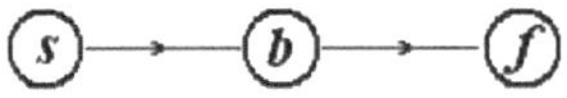

Figure 3.15. Illustration of formula (3.2).

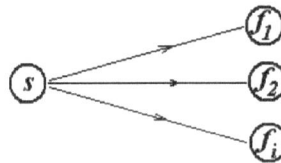

Figure 3.16. Illustration of formula (3.4).

But the very first rule, the rest are well known in the theory of probability as multiplication and addition theoremes of probabilities.

Unusual is only the first case; it follows out of it that the event's probability will be equal to the squared sum of amplitudes:

$$\langle f|s\rangle = \sum_i \langle f|s\rangle_i \Rightarrow |\langle f|s\rangle|^2 = \left|\sum_i \langle f|s\rangle_i\right|^2$$

Besides, it is clear that the squared sum will lead to the occurrence of co-multipliers corresponding to diffrent transition ways; this result is called «interference of amplitudes», *i.e.* there is a systemic description that focuses on the phenomenon without its strict «binding» to the rough spatial-temporal representation.

- *Now let us consider the above-mentioned experiments within a new description beginning with the first experiment - scattering on two slits. When a slit is closed, the probability of electron penetration from source I to point «x» of the detector will be (small letter p_i indicates the probability corresponding to histogram P_i in Fig. 3.1):*

$$\left|\langle x|\ I\rangle\right|^2 = \left|\langle x|\ 1\rangle\langle 1|\ I\rangle\right|^2 = p_1(x)$$

$$\left|\langle x|\ I\rangle\right|^2 = \left|\langle x|\ 2\rangle\langle 2|\ I\rangle\right|^2 = p_2(x)$$

(3.5)

At the same time, if both slits are open, we are to put it down as follows:

$$p(x) = \left|\langle x|\ 1\rangle\langle 1|\ I\rangle + \langle x|\ 2\rangle\langle 2|\ I\rangle\right|^2 = p_1(x) + p_2(x) +$$

$$+\langle x|\ 1\rangle\langle 1|\ I\rangle(\langle x|\ 2\rangle\langle 2|\ I\rangle)^* + (\langle x|\ 1\rangle\langle 1|\ I\rangle)^*\langle x|\ 2\rangle\langle 2|\ I\rangle$$

where complexly-conjugated values are marked with asterisks (according to determination of complex value's squared module. It is seen that it does not result in a simple sum of probabilities corresponding to histogram $P_3(x)$. The way of describing experiments with the quantum system we introduced correctly reflects experimental results. One can complicate the considered case, as it was in Feynman's lectures [16].

Let there be a source of photons S on the other side from the barrier. Thus, observation is divided into two stages. The first event is connected with radiation scattering on the barrier, the second one - on the photon. When the electron transmits through the slits, as a result of photon scattering on it, the corresponding detector will register quantum penetration. Herewith, we wll obtain some distribution of results and, moreover, we will know which way each electron transmitted (passed through) in the experiment. We build up our experiment so that it would be possible to outline an object's movement trajectory, to indicate the spatial area where it is certainly present from the source to the screen. Here it is important that electron localisation allows us to pass on to its object decsription.

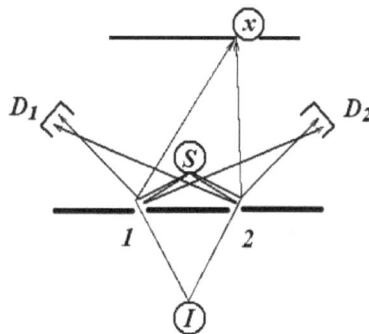

Figure 3.17. Experimental control of the ways.

First we will present the amplitudes for all possible outcomes which can be observed in such an experiment. Let, for instance, the photon wavelength is so big (more than interslit distance) that photon registration is posssible in both detector D_1 and D_2 under scattering on the electron at any slit. In the experiment, each electron dissipaton act will be *accompanied by the event which consists in the actuation of detector D_1 or D_2. The probability of the first one, using formulas (3.1) and (3.3), could be formulated as:*

$$|\langle xD_1 | SI \rangle|^2 = |(\langle x|1\rangle\langle 1|I\rangle\langle D_1|1\rangle\langle 1|S\rangle + \langle x|2\rangle\langle 2|I\rangle\langle D_1|2\rangle\langle 2|S\rangle)|^2 \tag{3.6}$$

and for the second event:

$$|\langle xD_2 | SI \rangle|^2 = |(\langle x|1\rangle\langle 1|I\rangle\langle D_2|1\rangle\langle 1|S\rangle + \langle x|2\rangle\langle 2|I\rangle\langle D_2|2\rangle\langle 2|S\rangle)|^2 \tag{3.7}$$

To simplify the expressions, let us introduce the following indications, considering the symmetry of our experiment:

$$\langle D_1|1\rangle\langle 1|S\rangle = \langle D_2|2\rangle\langle 2|S\rangle = \varphi_1 \text{ and for } \langle D_2|1\rangle\langle 1|S\rangle = \langle D_1|2\rangle\langle 2|S\rangle = \varphi_2$$
$$\text{and } \langle x|1\rangle\langle 1|I\rangle = \psi_1, \langle x|2\rangle\langle 2|I\rangle = \psi_2$$

Then expression (3.6) can be reformulated like this:

$$|\langle xD_1 | SI \rangle|^2 = |\psi_1\varphi_1 + \psi_2\varphi_2|^2$$

expression (3.7) like.

$$|\langle xD_2 | SI \rangle|^2 = |\psi_1\varphi_2 + \psi_2\varphi_1|^2$$

and

$$|\langle xD_1 | SI \rangle|^2 + |\langle xD_2 | SI \rangle|^2 = |\psi_1\varphi_1 + \psi_2\varphi_2|^2 + |\psi_1\varphi_2 + \psi_2\varphi_1|^2 =$$
$$= (|\psi_1|^2 + |\psi_2|^2)(|\varphi_1|^2 + |\varphi_2|^2) + (\psi_1\psi_2^* + \psi_1^*\psi_2)(\varphi_1\varphi_2^* + \varphi_1^*\varphi_2)$$

Considering the accepted indication (3.5), the total (sum) of probabilities will be equal to.

$$\sum_i |\langle xD_i | SI \rangle|^2 = (p_1(x) + p_2(x))(|\varphi_1|^2 + |\varphi_2|^2) + (\psi_1\psi_2^* + \tag{3.8}$$
$$+\psi_1^*\psi_2)(\varphi_1\varphi_2^* + \varphi_1^*\varphi_2)$$

and it is possible to suppose that $(|\varphi_1|^2 + |\varphi_2|^2) = 1$ is just the probability of observing the photon registration in any detector. Thus, the resultant probability of electron transfer from I → «x» has two summands, the first of which is the sum of scattering probability on slit 1 or 2, and the second one reflects the interferential origin (nature) of the phenomenon. It is possible to mention two extreme cases: the first one is complete indistinguishability of electron transmission way, which means $\varphi_1 = \varphi_2$. Then, obviously, we will get formula (3.1), $\langle f | s \rangle = \sum_i \langle f | s \rangle_i$ i.e. a complicated picture corresponding to distribution P(x) (Fig. (3.1) (a)) will be registered in the experiment.

At the decrease of wavelength, «controlling the way» of radiation, the photon scattering probability towards «alien» detector - φ_2 will tend to zero. In the second extreme case, we realise the precise control of transmission way, and then expression (3.8) transforms into distribution $P_3(x)$ that corresponds to classical addition of probabilities.

The introduced way of describing experimental data turns to be simple and convenient to reflect experimental results. Let us demonstrate this in the case of scattering of nuclei ^{4}He and ^{3}He. We analysed three examples in which, in the accepted designations, may be put down the following way (if different objects scatter and detectors register only particles of one sort): $|\psi(\theta)|^2 = |\langle D_1 | I_1 \rangle \langle D_2 | I_2 \rangle|^2$. If detectors register any particles, it is possible to formulate that full (complete) probability will be:

$$p + p' = |\psi(\theta)|^2 + |\psi(\pi - \theta)|^2 = \left| \langle D_1 | \; I_1 \rangle \langle D_2 | \; I_2 \rangle \right|^2 + \left| \langle D_2 | \; I_1 \rangle \langle D_1 | \; I_2 \rangle \right|^2$$

and, for angle $\theta = \pi/2$, we will obtain probability $2p$. And in the case when only ^{4}He nuclei scatter, we can never specify which source has emitted the particles registered by a certain detector. This variant corresponds to the addition of amplitudes, and it leads to the four-fold increase of probability $|\psi(\theta) + \psi(\pi - \theta)|^2$ for angle $\pi/2$! In variants 1 and 2, it is possible to describe the system using terms of scattering particles. The thing is not, if we extract the information from the experiment about what particle is registered by a certain detector or not. The main thing is that these are principally different situations realised in setting the experiment.

It is common to think that certain properties selection of an observed phenomenon is carried out in the experiment - either wave or particle. In fact, one variant of events is connected with the expression of the whole phenomenon, and it requires the corresponding description. In other case, object description is adequate. Note that wavelength reduction of «peeping» radiation will lead to a change of the analysed distribution. Such a change of the histogram, in the registration of a big number of particles, will carry superpositions of two distributions - $P(x)$ and $P_3(x)$.

Now let us pass on to the analysis of photon polarization. Remember, we had three filters successively located one by one at the rotation angle $45°$ towards the previous one. In these experiments, the result does not depend on filters initial orientation; so we can speak about the indefiniteness of the state of radiation polarization towards them before measurement.

- *Further on, with photon polarization kept in mind, we will also dwell upon its spin. The direction of wave electric field was chosen as the direction of radiation polarization; the term «spin» will be used for photon as radiation quantum. Generally, spin is the internal degree of freedom of a quantum system not connected with its spatial changes as a whole.*

  Any particle's spin may be set in space with three values - vector length accepted to be equal to 1 (Bloch sphere) and two angles, as shown in Fig. (3.18) (a). Then it will be demonstrated that spin may be used as qubit in quantum computations. In this case the state can be registered as:

  $$|\psi\rangle = \cos\frac{\theta}{2}|0\rangle + e^{i\varphi}\sin\frac{\theta}{2}|1\rangle$$

  Circumpherence is sufficient to present the photon spin (Fig. 3.18 (b)). Here, the situation may be explained in the following way. Imagine the photon with an impulse along direction «z». Then the photon spin has only «x» and «y» components which may be set by only one angle towards arbitrarily chosen axes in the plane perpendicular to its impulse.

  Next, it will make clear that, to present the properties of qubit, a 2D basis is required in the Hilbert space. Experimentally, quantum operations were realised with both photons and spins of different quantum systems, which is suggestive of their equivalents as qubits.

The photon spin may be put down as decomposition over states related to two mutually perpendicular directions (Fig. (3.3)), such as, $|\psi\rangle = a\langle \leftrightarrow | I \rangle + b\langle \updownarrow | I \rangle$, where a and b - coefficients, $\langle \leftrightarrow |$ and $\langle \updownarrow |$ - bra-vectors. As we are only interested in the vector direction, its value may be accepted single; besides, suppose

$|a|^2 + |b|^2 = 1$. generally, a and b may be complex numbers. Note that the choice of orthogonal basis is fully arbitrary: any two single orthogonal vectors may be chosen as basis. After passing through the first filter, we will obtain the state of trasmitted photons $0\langle \leftrightarrow | I \rangle + 1\langle \updownarrow | I \rangle$. It is not surprising that, when setting the second filter perpendicularly to the first one, the photons detection probability on the screen is equal to zero: $| \psi |^2 = | 0\langle \leftrightarrow | \leftrightarrow \rangle + 1\langle \leftrightarrow | \updownarrow \rangle |^2$. Earlier, we settled to think the scalar product of bra and cket of one and the same vector to be equal to 1, the product of orthogonal ones being equal to zero. Thus, the first summand turns to zero because of the zero coefficient at the vectors product and, in the second summand, the product of vectors turns to zero.

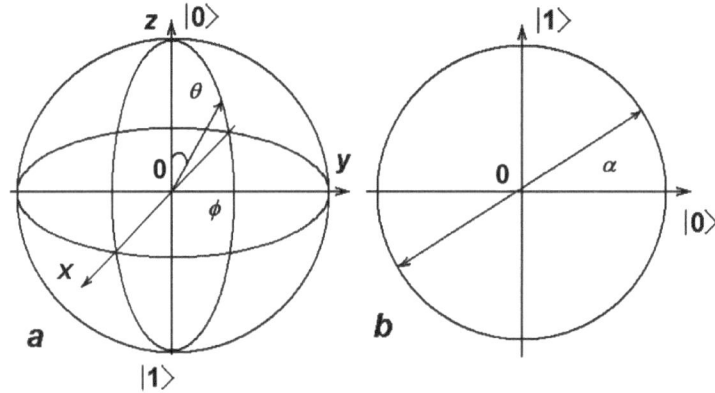

Figure 3.18. Bloch sphere (a), photon spin (b).

- *It is known from geometry that, with the rotation of co-ordinates system at an arbitrary angle α, the vector components will transform on the following formulae:*

$$x' = x \cos \alpha + y \sin \alpha$$
$$y' = -x \sin \alpha + y \cos \alpha$$

Setting (installation) of the intermediate filter turned at $45°$ enables us to present the spin of the photon that passed through the first filter towards the next one as:

$$| \nearrow \rangle = \cos \alpha | \leftrightarrow \rangle + \sin \alpha | \updownarrow \rangle \text{ or } | \nearrow \rangle = \frac{1}{\sqrt{2}} | \leftrightarrow \rangle + \frac{1}{\sqrt{2}} | \updownarrow \rangle \quad | \searrow \rangle = -\sin \alpha | \leftrightarrow \rangle + \cos \alpha | \updownarrow \rangle$$

$$\text{or } | \searrow \rangle = -\frac{1}{\sqrt{2}} | \leftrightarrow \rangle + \frac{1}{\sqrt{2}} | \updownarrow \rangle$$

And, for the orientation of the intermediate filter chosen in the experiment, the wave transition function will be

$\psi = \langle \nearrow | \updownarrow \rangle = \left(\frac{1}{\sqrt{2}} \langle \leftrightarrow | + \frac{1}{\sqrt{2}} \langle \updownarrow | \right) | \updownarrow \rangle = \frac{1}{\sqrt{2}}$. Now, it is becoming clear that the photon flow intensity, proportional to the square of probability amplitude will fall down twice behind this polarizer. On the full analogy, it is possible to present the photon that transmitted in the basis of the last polarizer turned (rotated) at the same angle towards it. As a result, radiation intensity on the screen will be eight times lower than the initial.

Thus, we matched the symbolic description and its application rules that allow us to describe phenomena in the field of quantum mechanics being completely adequate to experimental data.

Besides, we established the thing that there are two types that describe our representations - wholistic and object. Therefore, we got rid of the superposition principle and dualism of the notion «wave - particle», having preserved the concept of «particle» only for object presentation. It is also clear from the analysed examples that properties of quantum systems, before being measured in certain conditions (some basis), are indefinite (non-determined).

- *It is also necessary to emphasise the fact that, when formulating experimental results in quantum mechanics, we had to introduce the notion of «probability amplitude» which may be built up in the Hilbert space. To make the description of phenomena - within the space - and the space of object properties not contradictious, the laws of both an object (or a system) and its total (summed) characteristics should be observed. It leads to a close connection of two presentations. It is worth emphasizing the statement formulated by Neumann in his contribution «Mathematical bases of quantum mechanics» [12] about the thing that linear operator F, being functional in the Hilbert space, corresponds to value F. Self-conjugated operators correspond to real values, and the corresponding operator functions - to functions of values. Thus, already in 1932, Neumann formulated his statement about the thing that all measurable properties of objects (physical properties were meant) are described in the Hilbert space. It is relevant!*

The Hilbert space proper is the space of functions and, first, it was introduced as a space of sequences with a convergent set of squares (the so-called «l_2 space»). Elements (vectors) of such a space are infinite (endless) numeric sequences $x = (x_1, x_2, ...)$, such ones when the set $(x_1^2 + x_2^2 + ...$ converges. The sum of two vectors $\vec{x} + \vec{y}$ and vector $\lambda \vec{x}$, where λ - a real number, is determined as follows:

$$\vec{x} + \vec{y} = (x_1 + y_1, x_2 + y_2, ...)$$
$$\lambda \vec{x} = (\lambda x_1, \lambda x_2, ...)$$

For any vectors $\vec{x}, \vec{y} \in l_2$, the formula $(\vec{x}\,\vec{y}) = (x_1 y_1, x_2 y_2, ...)$ determines the scalar product, and a non-negative number is understood as the length (norm) of vector x:
$$\| \vec{x} \| = \sqrt{(\vec{x}\,\vec{x})} = \sqrt{x_1^2 + x_2^2 + ...}$$

Scalar product is always finite and satisfies the inequality: $|(\vec{x}\vec{y})| \leq \| \vec{x} \| \| \vec{y} \|$

Space L_2 [a, b] of all measurable functions, set on the segment (span, section) [a, b], having their finite integrals $\int_a^b f^2(x)dx$, is another example of the Hilbert space. Any functions $\varphi_i(x)$ from l_2, having the property of orthogonality $\int_a^b \varphi_i(x)\varphi_j(x)dx = 0$ and normalisation $\int_a^b \varphi_i(x)^2 dx = 1$, may function as single vectors here. The presentation of vectors (signals decomposition) as Fourier-representation may be a good example.

In a broader sense, the arbitrary endlessly measured vector space (real or complex) - in which scalar product is set and which is complete, related to its norm initiated by this scalar product, - is understood as Hilbert space. It is the complex vector space that is used in quantum physics. Sometimes, the Hilbert space has finite dimensionality. In case of $\frac{1}{2}$ spin, it is two-dimensional, and its elements are linear combinations of two states $|\uparrow\rangle$ and $|\downarrow\rangle$.

It is necessary to note that, also, other kinds of spaces are used in physical representations. For instance, Minkowski space, multi-dimensional dislayered Kaluza-Klein spaces in the superstrings theory, etc., are considered in the special theory of relativity.

MENTAL MAP AND SPATIAL-TEMPORAL RELATIONS

Object-based modeling of reality, first and second signal systems, metric of mental space and special theory of relativity.

To understand what we describe, it is necessary to clear out the way we make it. Until processing, the algorithms of data we obtain are «hacked», it is not possible to realise the reason for paradoxes in our theories, including the quantum theory. Setting the perceprion program proceeds for an individual's lifetime and is connected with the evolution of the whole genus *Homo sapiens*. It is possible to trace how meaningful the recognition of regularities of making up perspectives has been in visual perception in the history of civilisation. This process is first reflected in painting, then in architecture and, finally, it was formulated with the language of mathematics. Just compare Egyptian murals and pictures of Renaissance. Man has found the way to present the world vision to self and the surroundings. The necessity for communication, exchange of information is one of the basicic needs of social human. Not depending on the thing if we realise it or not, we exist in one unified informational space that has been created for the whole period of mankind's evolution. Let us focus on the point in more detail.

To have a possibility to compare senses and model (represent) «outer Reality» to an individual, it is necessary to construct some mental space [23], to set its *metrics* and *coordinate system* [24]. The latter means determining «neutral» or indiffrent levels of senses («zero points (dots)») and setting primary oppositions on the type «favourable - harmful», in relation to which the evaluation of intensity and the sign of senses proceeds. The nature created this inventory having not taken care of our recognising it. In general, it is also possible to consider mental spaces evolution in connection with the evolution of nervous sytem and receptive apparatus that allow us to make various comparisons - from elementary to higher, extended forms; mathemeticians have partially made it when investigating spaces of different types [25].

At the next stage, it is possible to realise *object modelling* of Reality, pointing out mental space dots as «complexes of senses» (feelings) which are relevant for an individual's survival, *i.e.* to build a *mental map*. Actually, *mental map* is a limited semiotic model of Reality «on-the-other-side» of senses («World mental model» [26]) that allows an individual to orient him or herself in it and to plan (program) one's actions aimed at satisfying burning needs. The adequacy of this model is checked or revealed by the efficiency of needs satisfaction, *i.e.* «survivability», welfare, *etc.* In plain words, the way of perceptual world division into some conditioned units turned to be most efficient for the solution of problems in real time regime. Subsequently, with the development of language, these units outlined as objects in the mental space and their description was developing according to certain objective regularities to satisfy various needs, including that of communication.

Obviously, objects *due to their way of construction and structure*, already make *sense* for an individual, since they automatically correlated with the individual's needs. In the other words, the «external» signals [25] are sorted in relation to *interceptive signals* (feeling of hunger, thirst, pain, *etc.* which are independent from our arbitrary will, but referred to us as «internal medium», *i.e.* body, which reflects the natural needs of our organisms) and, actually, they satisfy these neens, either direct or indirect ones. Correlation of objects on the mental map allows us to make up already secondary, more *abstract relations* and oppositions that determine *the signs of the next level* (digit, bit) describing an *object situation, operations and processes.*

---

[23] Space is a logically thinkable form (or structure) that serves as medium in which other forms and constructions are realised. In modern mathematics space is defined as a multitude of some objects which are called its dots; these are any geometrical figures, functions, states. Considering their multitude as space, only those properties of their totality accepted on the definition are analysed. Relations among dots multitudes determine space geometry. The basic properties of these relations are expressed in the corresponding axioms. In fact, space determines (within given qualities) the possibilities of constructing these or other objects, setting certain processes and states.

[24] Depending on an individual psychical state.

[25] Signal as an informative part of the process, having sense, relevant for and registered by an individual.

But, anyway, all «signs» of signal system I - in their origin - are not meant for *bilateral communication*, and our mind's mental map is not available for the direct perception by other individuals. If we strictly approach to the definition of sign, then it is an *object communication means*, and «feeling» is just a *subjective marker* of a constitutive element of «irritation» (disturbance) which is relevent for an individual.

Full signs based on the so-called «external object means,» implicitly presented on mental maps of other individuals, - *i.e.* signs available to the perception by *others*, - are necessary for interindividual communications (individual ↔ individual). Actually, speech originates as «naming» of relevant points that determine objects on the mental map, by associations of certain external processes - realised by an individual in one's surrounding (speech, letter, *etc.*) - with them. Other individuals interpret them as *signals*, and they are again transformed into «complexes of senses» - *signs of the second type*, - as they are not «non-conditionally» natural and they are conditioned by «social agreement» [26]. They are initially «bound» with *signal* irritants (stimuli) and are a means of *communication.* Note that, in the *sign-name*, its co-ordinate nature and notional component turn to be implicit [27]. In its structure, everything which is behind the sign is always limited [28] and is bound to some coordinate system and burning needs, and «name» makes an illusion of absolute «self-identity» of an object (independence of its meaning from the reference sytem and needs). For instance, a change of psychical state (transfer to the state of alarm) inevitably changes the «indifferent point» that sets the origin of co-ordinates and makes us outpoint the signals of other type or intensity and interpret them in some other way. This adaptive change of countdown system will, surely, affect an *object's absolute meaning*.

Thus, with the experimental modelling of a «named» object, as a sign of the second level, in *space of properties*, we will inevitably get its *various co-ordinate representations* for different individuals being in different psychical states. We will have different lexical meanings of the word-name in language, or an object's states which, actually, characterise individual object perceprion in different psychical states or situations [29].

As the ethnos that creates the language usually lives in more or less standard conditions and belongs to one biological species, then also interindividual commonality and stability of lexical meanings in this language follow from the objective generic similarity of human mental maps. Naturally, for the development of common (lexical) meaning of words, similar, *most probable states for the population* are pointed out, - otherwise the very possibility of interindividual communication is lost.

Another problem of psychosemantics is definition of the word's meaning (sense, gist). As an object is described by the totatlity of properties having a diffrent value for *various needs satisfaction*, then it is obvious that, when they are actualised and satisfied, *the sense (purpose) of an object* (what an individual needs it for and in what

---

[26]This aspect of speech generation is quite widely analysed in scientific literature (see, *e.g.*, [27].

[27] Signs of this type are full. They may be used to describe objects as points (dots) indicating their coordinates in some semantic sub-space, *i.e.* to realise «second-signal» simulation of «first-signal» mental map of an individual (Suprun A., Yanova N., Nosov K., 2007). Naturally, these points may be further named as A, B, C, *etc.* However, metric traits and «co-ordinate nature» are still implicitly present in their meanings: *e.g.* the words «apple» and «pear» are semantically closer to us than «apple» and «camel». Note that co-ordinate representation of a named object on the mental map is often not obvious, as it is constructed first in the semiotic representation of signal system I, and naming is realised in signal system II (after I.P. Pavlov). However, when modelling a mental map in the semantic space, we already have to realise the presentation of named objects explicitly, and «semiotic roots» of a name become obvious.

[28] The borderline implies the indication of a quantitative threshold within some property. Measure is necessary to determine quantity. Even when we compare abstract numbers, we still imply that comparison is made within one measure. We really do not think that 3 hours are less than 15 minutes and then, moreover, 15 hectars. Object borders (limits) are determined by quantitative expressivity of its properties intensity (*i.e.* within some qualities).

[29] *E.g.*, expressions «Morning star» and «Evening star» are different in their meaning, but they denote one and the same planet - Venus.

conditions), will also change [30]. But sense is already a more individualised and mobile characteristics of a word. It is reflected not in the absolute determining it in the space of properties (*i.e. in meaning*), but in the *preferences rate of properties that compose* it in different conditions and states. It is obvious that a change of properties value leads to a change of preferences of various objects in a totality of some class [31], and it may be laid ground for the construction of motivation vectors in the semantic space.

Thus, mental map's psychosemantic simulation implies the following steps:

1. Making up the semantic space *adequate* to that of an individual's mental map (or present mentality) that allows us to describe objects of any type [32].

2. Definition of object as «superposition» of its possible meanings and probabilities of their actualisation (*i.e.* considering all of its most probable interpretations in the perception by present mentality).

3. Making up motivation space based on the semantic one and defining individual senses for objects stable (set) meanings.

Let us try to analyse the process of an object properties transformation perceived by our «computing system». Say, we are in a regularly moving system, *e.g.* train. We normally have two cups of water. Everything is the same - mass of water, volume (extensive physical traits), temperature, speed - related to the table and ground, - taste, colour (intensive traits), *etc.* Now we put a spoon of instant coffee and sugar into each cup. Then the mass of each cup will increase, the colour of water will change, so will its taste. But the content of both cups will still be identical in their properties. We are not focused now on the error in determining these properties. Let us think it to be small. If we take and mix the content of both cups, we will form a new object and, what is more interesting, such properties as temperature, speed, colour, taste have not changed their meaning (value) (intensive traits), whereas mass has increased twice. Now let us translate it into the language of mathematics.

At the first step of determining mental map's model equivalent, *mental space reflection of « signal system I»* into *the vector space of properties of «signal system II»* seems to be most obvious and natural, which is, essentially, made by linguists - implicitly or explicitly - in a situation of direct [33] (absolute) determination of objects through their properties: $\Omega \rightarrow \vec{U} = U(q_1, q_2, ... q_n)$. As it was mentioned earlier, herewith, object Ω of mental space reflects into vector $\vec{U}$ in the space of properties $\{q_1, q_2, ... q_n\}$. It is easy to find the inadequacy of this reflection with the above-mentioned example: we poured two portions of coffee (Ω' and Ω'') together. In objects vector addition, coordinates of properties are to be summed, *i.e.* if $\Omega' \rightarrow \vec{U}'; \Omega'' \rightarrow \vec{U}''$ and $\Omega = \Omega' \cup \Omega'' \rightarrow \vec{U}$. Then correlation (ratio) $\vec{U} = \vec{U}_1 + \vec{U}_2 = U(q_1' + q_1'', q_2' + q_2'', ..., q_l' + q_l'')$ should be observed, adequately describing the «total» (summed) object Ω. However, neither taste (q_1) nor smell (q_2), nor speed (q_3) changed, related to the ground: $q_{1'} = q_{1''} = q_1; q_{2'} = q_{2''} = q_2; q_{3'} = q_{3''} = q_3$; though the object's mass (q_H) and the characteristics connected with it (volume, weight, *etc.*) are really added (summed) on the rules of vector space. It is

---

[30] *E.g.*: clothes may be not only a means of physical needs satisfaction of temperature comfort, but a means of higher social status, the sign of status, etc, and it is determined by its various properties being unequal for various needs satisfaction.

[31] *E.g.*: the rate of objects in the semantic class «clothes» - shirt, sweater, coat (their meanings unchanged) will be different under «heat» and «cold».

[32] It is still difficult to imagine that setting objects of a different type requires different principles of their mental maps construction. We are not going to discuss here the absurdity of this supposition; we will just confine it to Okkama principle of savings: *do not introduce excessive entities*

[33] Indirect (relative) definition of an object is realised in its correlation with other objects using metonymy and metaphor. Absolute definition is presented in the enumeration of an object's properties and their intensities (it is common in dictionaries).

possible only in case if, in our representation, both *angular* and *linear* coordinates are simultaneously used. In fact, the difference is conditioned by the possibility of introducing a limited normed basis in one case and that of the spatial norm - in the other. It is obvious that angular coordinates of vector (q_1, q_2, q_3), that determine its direction, do not change when its length is doubled. But its length, which we have to correlate with mass q_H, will be really doubled.

Hence, to make reflection $\Omega = \Omega' \cup \Omega'' \to \vec{U}$ adequate to the mental map, even when describing *one property* q_i, we will need not *a one-dimensional space*, but *a plane* [34]. First, let us make the reflection of angular value into linear: $q_i \to \vec{V_i} = C_{(i)} \cos \varphi_i$ ($C_{(i)}$ - a constant that determines the measure (scale) of the present property's variability) and then pass on to the formulation:

$$\vec{U}_i = \left\{ |\vec{U}_i| \cos \varphi_i ; |\vec{U}_i| \sin \varphi_i \right\} = \left\{ |\vec{U}_i| v_i; |\vec{U}_i| v_{Hi} \right\} = |\vec{U}_i| (v_i \vec{e}_i + v_{Hi} \vec{e}_H) = \{ U_{Vi}; U_H \}$$

as the angle may be determined by a space of not less than two dimensions. Herewith, $\vec{e}_i$ and $\vec{e}_H$ - single vectors, $|\vec{U}_i|$ - length of vector projection $\vec{U}$ onto i-property plane: $\vec{e}_i \times \vec{e}_H$ (see Fig. 3.19) expressed through «linear» component U_i (rigidness of property) and its «angular» component φ_i (intensity of property). It is obvious that:

$$\cos \varphi_i = V_i / C_{(i)} = v_i; \to v_{H_i} = \sin \varphi_i = \sqrt{1 - \cos \varphi_i^2} = \sqrt{1 - v_i^2} = \sqrt{1 - \frac{V_i^2}{C_{(i)}^2}}$$

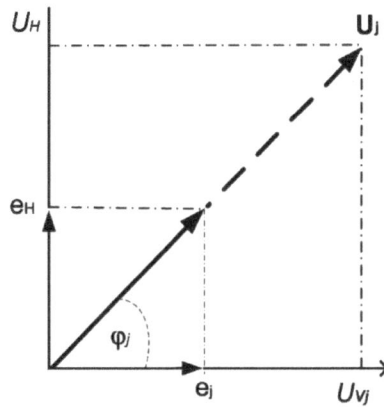

Figure 3.19. Object representation on the mental map.

Thus, the smaller angle φ_i is - and, respectively, the bigger the cosinus is - the higher the present quality's intensity will be. It implies the thing that the introduced scale is limited and non-linear. Close to zero, intensities are well approximated by linear dependence; however, as they approach to 1 - the present quality's extreme point - the changes become considerably non-linear. Practically, an «ideal» in the semantic space - it is only there that properties analysis is realised - cannot be achieved. Hence, in general, qualties intensities may be put down as follows:

$$V_i = C_{(i)} \cos \varphi_i = C_{(i)} \left(\frac{U_i}{\sqrt{U_i^2 + U_H^2}} \right) = C_{(i)} \frac{U_i}{|\vec{U}_i|}$$

---

[34]As setting an angle is possible only in plane (flat) continuum.

where $C_{(i)}$ - some constants or scaled coefficients depending on the chosen sytem of properties measurement units.

Thus, for an adequate semiotic reflection of properties in language, it is necessary not only to indicate its *quality* and *intensity* (set angle φ_i), but also *rigidness* [35] (length of vector U_i), i.e. *property's sustainability to a change of its intensity*. It is this object's representation that may be called *semantic*. Under such a definition, all properties intensities become *limited* [36] because $-1 \le \cos\varphi_i \le 1$.

- *Let us comment on the concept rigidness (sustainability) with the example of colour perception. As it is known, all colours having this or other shade of colour are called chromatic. If we add white colour to the spectral light, we will make colours of one tone with a degree of their dilution with white light. The wavelength of spectral colour which, when diluted, results in the present colour, is called colour tone. Dilution degree is characterised by a parameter p which is called colour purity. Colour purity is normed per unit or 100%. Spectral colours have their colour frequency equal to 100%. However, our eye can differentiate only limited colours number of a similar colour tone. This number varies from 4 (yellow) to 25 (red) for the spectral colours of different colour tone. Thus, during perception, yellow is more rigid, when diluted, than red.*

Therefore, for a full object representation through a set of its properties on the mental map, it is necessary to use the space having axes of properties rigidness and intensity. Intensity of a property is normed for a quantitative comparison of objects on the degree of a property'e expressivity. Besides, the used description should have the characteristics of additivity, in the sense of its ability to include new or necessary properties by adding the corresponding axes. It is known from experience that such a possibility does not lead to reconsideration of the whole description each time when new circumstances occur.

We know it experimentally that the object has «residual» rigidness also at $U_{Vi} = 0$ (actually, we cannot «decompose» it into properties. Furthermore, not all of them are currently known). In the present case, it is obvious that: $\vec{U}_i = \vec{U}_H$ (see Fig. (**3.19**)). Herewith: $\varphi_i = \pi/2, v_i = \cos\pi/2 = 0$. When fixing value U_H for the present object as «zero» or initial rigidness of the object's property i.e. its length in the reference system in which $v_i = 0$, and thinking it to be an *unchangeable characteristic of the whole object* in its «own» reference system, we will get the result of the thing that, in a general case, the object's property is adequately reflected in the semantic space as a linear one-dimensional value:

$$|\vec{U}_i| = \frac{U_H}{sin\varphi_i} = \frac{U_H}{\sqrt{1-\cos^2\varphi_i}} = \frac{U_H}{\sqrt{1-\dfrac{V_i^2}{C_{(i)}^2}}}$$

Then

$$U_{Vi} = |\vec{U}_i|\cos\varphi_i = \frac{U_H}{\sqrt{1-\dfrac{V_i^2}{C_{(i)}^2}}}v_i = \frac{U_H/C_{(i)}}{\sqrt{1-\dfrac{V_i^2}{C_{(i)}^2}}}V_i = \frac{M_{0(i)}}{\sqrt{1-\dfrac{V_i^2}{C_{(i)}^2}}}V_i = M_{(i)}V_i = P_i$$

which coincides with relativistic *impulse description of a physical object* (in case of the only property V). Here (in accordance with physical terminology) M_0 - rest mass, M - full mass of object, P - impulse). In

---

[35] True, in physics and, according to the definition, mass is *stability* (sustainability) of such a property as speed to the force impact aimed at its change. And sustainability (stability, resistance) to changes is rigidness proper.

[36] E.g. terminal (limit, extreme) mechanical velocity C in physics or introversion in psychology, *etc.* By the way, this limitation is connected not with «laws of nature» but the character of properties representation on the mental map of a subject in any reference system.

such a representation, it is easy to make a mistake and to think *property rigidness* to be a separate independent property which takes place in psychology (just as in certain physical theories, *e.g.* theory of gravitation). Actually, it turns out that, in physics, a transition from *velocities* to *impulses* means transition *from non-complete space of properties to the semantic space-* only in which *semantic components of interacting objects* [37] *description are preserved - adequate to mental map.*

As we accept rigidness U_H as a stable «object characteristic» at all its changes (and, thus, we «painlessly» fix one degree of freedom), then, in general, we can write down an identity for any object and an arbitrary property: $U_H^2 = U_i^2 - U_{Vi}^2$ or.

$$1 \equiv \frac{U_i^2}{U_H^2} - \frac{U_{Vi}^2}{U_H^2}.$$

Formally, the last identity may be reformulated as: $\cosh^2\theta - \sinh^2\theta = 1$.

Here, $\cosh\theta = U_i / U_H$ - hyperbolic cosinus, and $\sinh\theta = U_{Vi} / U_H$ - hyperbolic sinus. It follows from here that:

$\tanh\theta = \sinh\theta / \cosh\theta = U_{Vi} / U_i$ or $\tanh\theta = U_i \cos\varphi / U_i = \cos\varphi = V / C$. Hence,

$$V / C = \tanh\theta = \tanh(\theta_1 + \theta_2) = \frac{\tanh\theta_1 + \tanh\theta_2}{1 + \tanh\theta_1 \tanh\theta_2} = \frac{V_1 / C + V_2 / C}{1 + \dfrac{V_1 V_2}{C^2}}$$

where we obtain the rule of properties addition from:

$$V = \frac{V_1 + V_2}{1 + \dfrac{V_1 V_2}{C^2}} \tag{3.9}$$

analogous to that of special theory of relativity. It would be strange to expect that nature has been developing some special algorithms and rules during evolution for the presentation of «physical» feelings (properties). However, comparison with physical representations is first convenient in connection with a more developed mathematical apparatus of this discipline. It is clear that *the space of properties* is *pseudo-Euclidean* from *the limitation of properties variability area*, whence relativistic regularities (effects) follow [38].

From equality $U_i^2 = U_{Vi}^2 + U_H^2$ it follows that:

$U_H^2 / sin^2\varphi = (U_H^2 \cos^2\varphi) / sin^2\varphi + U_H^2$ (let us confine it to one property and omit the index at φ and V), or

$$\frac{U_H^2}{\left(\sqrt{1 - \dfrac{V^2}{C^2}}\right)^2} = \frac{U_H^2 \dfrac{V^2}{C^2}}{\left(\sqrt{1 - \dfrac{V^2}{C^2}}\right)^2} + U_H^2$$

---

[37] It is obvious that, not considering mass, physicists will not begin to predict the consequences of bodies interaction, unlike psychologists, that easily take to the solution of such problems explaining all their misses with the complexity of a problem, but not with errors in its setting.

[38] It is based on two axioms: laws of nature are the same in all reference systems; maximal speed of interaction with Universe is limited and it is the same in all countdown systems.

The last equality is:

$$\frac{\dfrac{U_H^2}{C^2}C^2}{\left(\sqrt{1-\dfrac{v^2}{C^2}}\right)^2} = \frac{\dfrac{U_H^2}{C^2}V^2}{\left(\sqrt{1-\dfrac{V^2}{C^2}}\right)} + \frac{U_H^2}{C^2}C^2$$

which is equivalent to:

$$\frac{M_0^2 C^2}{\left(\sqrt{1-\dfrac{V^2}{C^2}}\right)^2} = \frac{M_0^2 V^2}{\left(\sqrt{1-\dfrac{V^2}{C^2}}\right)} + M_0^2\, C^2$$

It follows from here that $M^2 C^2 = P^2 - M_0^2 C^2$ or $(MC^2)^2 = C^2(P^2 - M_0^2 C^2)$, which is equivalent to Hamilton function in physical objects description:

$$E = C\sqrt{P^2 - M_0^2 C^2}\ .$$

Herein:

$$E = MC^2 = \frac{M_0 C^2}{\sqrt{1-\dfrac{V^2}{C^2}}} \tag{3.10}$$

E - objects's energy, M_0 - rest mass, P - impulse. The index is omitted in the formulas at φ and V as we analysed one concrete property.

In fact, we have already determined the situation so that *semantic laws* of «energy»- saving and «impulse» and Lorentz tranformations that may be used when describing objects of any origin (nature) in the semantic field, and those being equivalent to special relativistic laws in physics, univocally follow from it (situation). These are totally *not physical, but universal semantic laws, actually means, laws of conservation and transformations of a system of objects description in the semantic space.*

- *The rule of properties addition (3.9) precisely corresponds to the rule of velocities addition in relativistic mechanics and is realised in Minkowski [39] space. It is not difficult to check that, even the sum of extreme (limit) values C_j still results in a limit value.*

  At $V_j \ll C_j$, formula (3.9) becomes classical: $V_j = V_{j1} + V_{j2}$. However, if most part of everyday tasks is solved in physics using the classical case, in psychology and sociology, - in connection with lower rigidness of properties under study - it is, as a rule, obligatory to consider non-linearity, i.e. even a usual calculation of mean values may lead to big errors. Experimental evaluations of sports achievements, etc. [40] can be a good psychological example to demonstrate it. Subconsciously, our mentality perceives a continuous on-growth of difficulties on the way to perfection as a natural phenomenon: every next step becomes more difficult than the previous one, and differences in achievements are less noticed. Let ideal «C» is equal to one, and we determine the difference between intensities (0.2 and 0.1)

---

[39]This space determines the mental map's metric.

[40] Relativistic relations also allow us to explain Stevens law in psychophysics (see Suppl. C).

of some property of two objects. The difference will be: $(0.2-0.1)/(1-0.2\cdot0.1/1^2)=0.1/(1-0.02)\approx0.1$, *as it is common in classics. However, for the expressivity of properties 0.9 and 0.8, the real difference will be higher:* $(0.9-0.8)/(1-0.9\cdot0.8/1^2)=0.1/(1-0.72)\approx0.357$. *Therefor expenditures also [41] of ΔE (formula 3.12) for a change of property's intensity (e.g. a sports result, personal portrayal, etc.) will be much higher in the second case:*

$$\frac{\Delta E_2}{\Delta E_1}=\left(m\frac{1^2}{\sqrt{1-0.9^2}}-m\frac{1^2}{\sqrt{1-0.8^2}}\right)/\left(m\frac{1^2}{\sqrt{1-0.2^2}}-m\frac{1^2}{\sqrt{1-0.1^2}}\right)\approx$$

$$\approx\frac{0.627}{0.0256}\approx25.4$$

though mental differences in the perception of a property's intensity are equal in both cases 1 and 2: $(0.2-0.1=0.9-0.8)$.

Now we obtain the formulae for calculation of rigidness U_H and semantic co-ordinates of object Ω. As, $\cot\varphi_j=U_j/U_H=u_j/u_H=v_j/v_{Hj}$ where u_j and u_H - normed semantic co-ordinates, and v_j - normed intensities of properties:

$\vec{u}=\vec{U}/|\vec{U}|,v_j=V_j/C_j=\cos\varphi_j,v_{Hj}=V_{Hj}/C_j=\sin\varphi_j$, it follows from here that:

$$U_j=U_H(v_j/v_{Hj}) \tag{3.11}$$

As, $\vec{U}_2=\sum_{j=1}^n\vec{U}_j^2+\vec{U}_H^2=\sum_{j=1}^nU_j^2+U_H^2$ and then, using ratio (3.11), we will get:

$$\vec{U}^2=\sum_{j=1}^nU_H^2\frac{v_j^2}{v_{Hj}^2}+U_H^2=U_H^2\left[\left(\sum_{j=1}^n\frac{v_j^2}{v_{Hj}^2}\right)+1\right]=U_H^2\left[\left(\sum_{j=1}^n\frac{v_j^2}{1-v_j^2}\right)+1\right]$$

and it follows from here that:

$$U_H=\frac{|\vec{U}|}{\sqrt{\left(\sum_{j=1}^n\frac{v_j^2}{1-v_j^2}\right)+1}}$$

Hence:

$$u_H=\frac{1}{\sqrt{\left(\sum_{j=1}^n\frac{v_i^2}{1-v_j^2}\right)+1}} \tag{3.12}$$

$$u_j=u_H\frac{v_j}{v_{Hj}}=u_H\frac{v_j}{\sqrt{1-v_j^2}}=u_H\frac{V_j/C_j}{\sqrt{1-\frac{V_j^2}{C_j^2}}}$$

Then the absolute semantic co-ordinates U_j may be calculated on the formula:

---

[41]Expenses (expenditures) are proportional to energy.

$$U_j = \frac{V_j \, u_H \dfrac{|\vec{U}|}{C_j}}{\sqrt{1-\dfrac{V_j^2}{C_j}}} = \frac{V_j \, m_j}{\sqrt{1-\dfrac{V_j^2}{C_j^2}}} = M_j \, V_j \qquad\qquad (3.13)$$

Herewith: m_j - zero rigidness [42], M_j - full rigidness of j-property,

$$m_j = u_H \cdot |\vec{U}|/C_j \,, \quad M_j = \frac{m_j}{\sqrt{1-\dfrac{V_j^2}{C_j^2}}} \, a \,. \qquad\qquad (3.14)$$

Thus, physical representations are a special case of semantic way of objects description (objects representation on the mental map). Absolutely, the obtained ratios are objective and common for mental maps of all individual *Homo sapiens*. However, they refer not to *Reality proper, but to the mental way of its representation.*

As an example, it is possible to give the analogies with paradoxes in cartography, when practical necessity makes us reflect the Globe in plane Euclidean projections, i.e, topographical maps. So far, our practical needs have been limited by small territories within which our perception regularities have been developing, there have been no paradoxes. But, obviously, without distortions that grow towards the Earth poles, it is impossible to make up a wholistic «plane» map of all the world. Naturally, topological and metric distortions on a plane map can be expressed as a *precise «objective» law*. However, it refers not to objective reality proper, but to the *way of its representation.*

Thus, limitation by extreme (limit) values [43] ($\pm C_j$) of all properties on their intensity is just a consequence of a certain *way of objects representation on the mental map*, and it makes it finite and «observable» for the subject.

Let us additionally analyse the category of rigidness and its connection with other concepts. According to its definition, it reflects stability of an object's properties regarding the quantitative change of their intensities. Now let us find out the meaning of V_{Hj}. It is not difficult to get the addition formula for the value V_{Hj} we introduced:

$$V_{Hj}^2 = C_j^2 - V_j^2 = C_j^2 - \left(\frac{V_{j_1}+V_{j_2}}{1+\dfrac{V_{j_1}V_{j_2}}{C_j^2}}\right)^2 = \frac{(C_j-V_{j_1})^2 - V_{j_1}^2\left(1-\dfrac{V_{j_1}^2}{C_j^2}\right)}{\left(1+\dfrac{V_{j_1}V_{j_2}}{C_j^2}\right)^2} = \frac{V_{Hj_1}^2 \, V_{Hj_2}^2}{C_j^2\left(1+\dfrac{V_{j_1}V_{j_2}}{C_j^2}\right)^2}$$

$$V_{Hj} = \frac{V_{Hj_1} \, V_{Hj_2}}{C_j\left(1+\dfrac{V_{j_1}V_{j_2}}{C_j^2}\right)} = \frac{V_{Hj_1}\sqrt{1-\dfrac{V_{j_2}^2}{C_j^2}}}{1+\dfrac{V_{j_1}V_{j_2}}{C_j^2}} \qquad\qquad (3.15)$$

---

[42]Rigidness at a property's zero intensity.

[43] It explains, in particular, the reason for velocities limitation in physics by light velocity and strange paradoxes of special relativity theory. We would like to emphasise once more the thing that the reason for it is not physical, but psychological.

Each individual is described by a set of one's typical qualities and their intensities: force, balance, mobility of nervous processes, temperament, character, *etc.* Actually, an individual is a *certain system of references* out of which «external» processes are perceived and evaluated. It is obvious that, due to individual differences, *all external phenomena will be differently perceived from different countdown systems* [44]. Note that the psychological system of references is defined by the expressivity of psychological properties, the physical is being expressed by physical properties.

One can continue the check of this interpretation for its accordance with various scientific statements. For *e.g.*, it is possible to get the equation for Doppler effect by considering the description of anti-particle's properties, *etc.* It will be demonstrated in the chapter about the development of objects representation on the mental map. Now let us make some conclusions.

We have shown that an object's adequate description through its properties, in any theory, is possible only in the semantic space and requires obligatory indication of *not only properties intensities, but also their rigidness.* It is connected with the thing that properties are presented not in one-dimensional $\vec{e}_j$, but with plane continuum $(\vec{e}_j \times \vec{e}_H)$ in it, as it was mentioned above.

Thus, the chosen psychosemiotic way of objects description (objects representation on the mental map) leads to a full agreement with the physical reality outlook. Absolutely, the obtained ratios are *objective and common* for mental maps of all individuals of genus *Homo sapiens.* However, they are not related to Reality proper, but to the mental way of its representation.

It is possible to check relativistic correlations in the field of psychology using psychophysical material, as it is the only field where psychologists make quite precise measurements (see **Appendix C**). An extended algorithm of making a semantic space is presented in **Appendix B**.

For the correct perception of material on quantum phenomena, it is necessary to understand that the above-described way of object representation is not applicable in this case. At this very cognition level, the nature demonstrates the wholiness which cannot be adequately described in a usual paradigm. Accoding R. Feynman nobody understands quantum mechanics.

MATRIX DESCRIPTION OF QUANTUM SYSTEMS

Description of a quantum system's state with the «language» of linear algebra, definition of qubit.

In this place we will continue to specify the concepts introduced earlier in the description of quantum phenomena. It turns out that *bra* and *cket*-vectors may be presented within linear algebra as a line $\langle x|$ - bra and a column $|x\rangle$ - cket. In other words, vectors $\langle x|$ are said to form Ermitean-conjugated matrix-lines indicated as $\langle x| = |x\rangle^+$. Then, for a system having two different orthogonal and normed states, it is possible to point out a basis in the 2D complex vector space $\{|0\rangle, |1\rangle\}$. Any linear combination $|0\rangle$ and $|1\rangle$, including that with coefficients, may be presented in it as $a|0\rangle + b|1\rangle$. Note that the order of basis vectors is arbitrary. Let us choose basis vectors as, $|0\rangle = \begin{pmatrix} 1 & 0 \\ 0 & 0 \end{pmatrix} \equiv \begin{pmatrix} 1 \\ 0 \end{pmatrix}$ and $|1\rangle = \begin{pmatrix} 0 & 0 \\ 1 & 0 \end{pmatrix} \equiv \begin{pmatrix} 0 \\ 1 \end{pmatrix}$. Herewith, *bra*-vectors are transposed *cket*-vectors, *i.e.* $\langle 0| = \begin{pmatrix} 1 & 0 \\ 0 & 0 \end{pmatrix} \equiv \begin{pmatrix} 1 & 0 \end{pmatrix}$ and $\langle 1| = \begin{pmatrix} 0 & 0 \\ 1 & 0 \end{pmatrix} \equiv \begin{pmatrix} 0 & 1 \end{pmatrix}$. The combination

---

[44] Depending on how our personal qualities are expressed (anxiety, sociability, capability and others), we will evaluate other people. For instance, Russians think «imperturbability» to be a national feature of the Finnish, and «irascibility» - of Georgeans. Thus, even a choleric Finn, compared with a Georgean, will seem to be phlegmatic to a Russian person.

of *bra-* and *cket*-vectors, which was registered as $\langle x \| y \rangle$ or $\langle x \mid y \rangle$ just as indicating the scalar product of two vectors, may be obtained according to the rule of matrices multiplication. Element a_{ij}, being in *i*-line and *j*-column of the product, is equal to the following sum of products of multipliers element matrices: $a_{ij} = \sum_k b_{ik} c_{kj}$. In this case, it is obvious that $\langle 0 \mid 0 \rangle = \langle 1 \mid 1 \rangle = 1$ and $\langle 0 \mid 1 \rangle = \langle 1 \mid 0 \rangle = 0$.

- *Within linear algebra, the task of transforming vectors to other biases may be solved simply; it was analysed above. For instance, the transition from $|0\rangle$ to $|0'\rangle$, in the coordinate system's turned at angle α will be:*

$$|0'\rangle = |0\rangle \begin{pmatrix} \cos\alpha & -\sin\alpha \\ \sin\alpha & \cos\alpha \end{pmatrix} + |1\rangle \begin{pmatrix} \cos\alpha & -\sin\alpha \\ \sin\alpha & \cos\alpha \end{pmatrix} =$$

$$= \begin{pmatrix} 1 \\ 0 \end{pmatrix} \begin{pmatrix} \cos\alpha & -\sin\alpha \\ \sin\alpha & \cos\alpha \end{pmatrix} + \begin{pmatrix} 0 \\ 1 \end{pmatrix} \begin{pmatrix} \cos\alpha & -\sin\alpha \\ \sin\alpha & \cos\alpha \end{pmatrix}$$

$$|0'\rangle = \cos\alpha |0\rangle + \sin\alpha |1\rangle$$

Classically, individual states of n-particles, considered in the configuration space, are integrated using common multiplication. Configurational k-D space, with a number of measurements equal to that of a system's degrees of freedom and the temporal axis, is introduced for the conditional representation of the whole system at some point in this space. A change of the mechanical system's position and mutual position of its parts may be determined with generalised co-ordinates $\{q_1, q_2, ..., q_{k-1}, t\}$. If we consider these co-ordinates as k of Cartesian co-ordinates in the k-D space, then a certain point, called depicting, of this space will correspond to each configuration of this system.

Quantum states are integrated with tensor multiplication. We can feel the difference in the following example. Let there be a particle which may be characterised by two parameters, e.g. co-ordinate x at the moment of time t. Then, to describe it, a surface with the Cartesian co-ordinate system of two axes - where a particle's state may always be reflected with a point on the plane - set on it will be sufficient. If we add one more analogous particle, then, to reflect the state of two particles system with a point in space, we will have to increase the spatial dimensionality twice. True, in case of both particles temporal synchronisation, it is possible to exclude one axis, as it will be general (common) for them. In a general case, we will need a 6-D space for three particles, where n - space dimensionality - will be 2 × n. Such a representation is the description of the mechanical process. The situation is quite different for a quantum system when describing its properties that may not be presented univocally. Based on the possible number of variants of the observation's outcome, basis dimensionality in the Hilbert space is chosen. For instance, for n-systems having binary characteristics, we will get only 2^n variants. i.e., in a classical case, bases of vector spaces will be summed (added), or dimensionality of states space will linearly grow with a particle number: $\dim(X) + \dim(Y)$. For a quantum system, it is necessary to use the product. For example, tensor product of $\vec{X}$ and $\vec{Y}$ vectors has the basis $\{x_1 \otimes y_1, x_1 \otimes y_2, x_2 \otimes y_1, x_2 \otimes y_2\}$. Its order may be arbitrary. Generally, matrices A and B tensor product, sized $m \times n$ and $p \times q$, respectively, is expressed in the following way:

$$A \otimes B = \begin{bmatrix} a_{11}B & a_{12}B & ... & a_{1n}B \\ a_{21}B & a_{22}B & ... & a_{2n}B \\ ... & ... & ... & ... \\ a_{m1}B & a_{m2}B & ... & a_{mn}B \end{bmatrix}$$

and the product value will be $mp \times nq$.

Now it is possible to determine qubit - the basic concept for quantum computer. A two-level quantum system that carries out the superposition of the states was named qubit (quantum bit) by B. Schumacher [28] (Fig. (**3.18**)). Also the quantitative unit of quantum information called qubit, analogously to classical bit. Information-coding is realised by manipulation of separate qubits states and their totality. Qubit is a more general notion which also includes the classical bit, if to exclude quantum peculiarities of recording information. Moreover, qubits totality can develop intercepted states that do not exist in classical systems. They will be analysed further.

Note that, to simulate even not a big quantum system, an ordinary computer of huge power is needed, as modelling requires the preservation of data on an exponentially huge number of states. For instance, 2^{500} states will correspond to only 500 qubits, and this may be co-measured - in the order of magnitude - with the number of all atoms in the visible part of the Universe. The reason for the potentially big power of quantum computer is the possibility of using quantum states evolution as a computation mechanism. In plane words, the possibility of a quantum system to have several states allows it to disparallel the computing process in a quantum computer, that is, the space of phenomena has an immeasurably bigger content than that of the object. And it may lead to a considerable acceleration in results achievement using the correspondent computation algorithms.

- *Thus, e.g. for two qubits, the basis will be:*

$$\{|0\rangle\otimes|0\rangle,|0\rangle\otimes|1\rangle,|1\rangle\otimes|0\rangle,|1\rangle\otimes|1\rangle\}$$

and it may be registered in a concise form $\{|00\rangle,|01\rangle,|10\rangle,|11\rangle\}$. Three-qubit system's basis: $\{|000\rangle,|001\rangle,|010\rangle,|011\rangle,|100\rangle,|101\rangle,|110\rangle,|111\rangle\}$. There are 2^n basis vectors for the n-qubit system. Generally, we will put it as $|x\rangle$ to indicate $|b_n\, b_{n-1}\,...b_0\rangle$, where b_i - binary digit of number x. You can see that there is exponential space extension with an increasing particles number, as the product of vectors $\vec{X}$ and $\vec{Y}$ has dimensionality $\dim(X)\times\dim(Y)$.

If the same is expressed in the language of matrices, then tensor product of two qubits, which is going to determine a new basis $\{|00\rangle,|01\rangle,|10\rangle,|11\rangle\}$, will be:

$$|0\rangle\otimes|0\rangle=\begin{pmatrix}1\\0\end{pmatrix}\otimes\begin{pmatrix}1\\0\end{pmatrix}=\begin{pmatrix}1\\0\\0\\0\end{pmatrix};|0\rangle\otimes|1\rangle=\begin{pmatrix}1\\0\end{pmatrix}\otimes\begin{pmatrix}0\\1\end{pmatrix}=\begin{pmatrix}0\\1\\0\\0\end{pmatrix}$$

$$|1\rangle\otimes|0\rangle=\begin{pmatrix}0\\1\end{pmatrix}\otimes\begin{pmatrix}1\\0\end{pmatrix}=\begin{pmatrix}0\\0\\1\\0\end{pmatrix};|1\rangle\otimes|1\rangle=\begin{pmatrix}0\\1\end{pmatrix}\otimes\begin{pmatrix}0\\1\end{pmatrix}=\begin{pmatrix}0\\0\\0\\1\end{pmatrix}$$

It is clear, from the above-mentioned example, how space dimensionality increase and «mechanics» of this space's orths are reflected. It is relevant that not all states formed in the basis of two qubits may be decomposed over the states of initial qubits. For example, state $|00\rangle+|11\rangle$ may not be presented as a qubits product. In other words, we can not find such coefficients a_1, a_2, b_1, b_2 so that: $\left(a_1|0\rangle+b_1|1\rangle\right)\otimes\left(a_2|0\rangle+b_2|1\rangle\right)=c|00\rangle+d|11\rangle$, as:

$$\left(a_1|0\rangle+b_1|1\rangle\right)\otimes\left(a_2|0\rangle+b_2|1\rangle\right)=a_1a_2|00\rangle+a_1b_2|01\rangle+a_2b_1|10\rangle+b_1b_2|11\rangle$$

is also necessary for the first equality to have $a_1b_2 = 0, a_2b_1 = 0$ which cannot be without zeroing of the whole equality. The states that cannot be decomposed in the above-mentioned way are called intercepted. Such states present a situation which has no analogy in classical consideration, but they play a big role in the operation of quantum computer and quantum cryptography.

MEASUREMENT OF QUANTUM SYSTEM STATES

Successive measurement of two particles, renorming, entangled pairs, norming of probability and probability amplitude.

Let us consider the measurement of some quantum object's state, *i.e.* its transfer to the object representation in more detail. During quantum computer functioning, a computation cycle terminates in some state of the register that contains output data as qubits states. Now we will focus on the information read-out (decoding) process and on what occurs with the wave function of quantum phenomenon. Herewith, there are many complicated - not understood so far - problems. The thing is that equations of quantum mechanics are reversible in time which acts as a parameter. It is common to be used when calculating the observation probability of this or other property's value of a quantum phenomenon under consideration. This «time» allows us to calculate all possible states, just as it will be shown later with the example of a two-bit system. But it turns out that, finally, these states do not depend on time. Time, in this case, has no evolutionary meaning, and, depending on it, the system does not alter its properties. However, the use of this notion has settled (rooted down) in the theory and considerably complicates the understanding of its essence.

- *When measured, an irreversible change of a quantum system's state occurs, and that leads it to a final fixed result. Properly speaking, the thing that the process of measurement is beyond the consideration of quantum mechanics is acknowledged by all. It was not clear what it was conditioned by, what role an observer, one's mind plays on it - and if it does any. If an observer is «officially» not included in the scientific paradigm, all references to it are illegitimate.*

As it was mentioned above, when measured, a state is characterised by some probability of its realisation which, under single measurement, may not be determined. Part of the information which qubit embodies is lost in its measurement. Therefore, there raises a question as what information is it possible to obtain after the quantum computer program is over?

Measuring a quantum phenomenon leads to one of the system's states, which were based on the possibility of registering a required property. In other words, it is common to state the thing that measuring a quantum system projects the system's initial state over the sub-space of states space compatible with the measured value. The projection amplitude is scaled so that the full state's vector will be single, *i.e.* remain normed. The probability of any of the possible values in the measurement's result is equal to the sum of squared values of probability amplitudes for all probability components that describe the given property which is equal to 1.

- *Let us consider the example of measuring a two-qubit system [29]. As we already know, all measurements are obtaining a state in the basis $\{|0\rangle, |1\rangle\}$. All the possible states of a two-qubit system may be expressed as $a|00\rangle + b|01\rangle + c|10\rangle + d|11\rangle$, where a, b, c, d are complex numbers, and the norming condition means that $|a|^2 + |b|^2 + |c|^2 + |d|^2 = 1$. As a result of measuring the first qubit with probability $\left(|a|^2 + |b|^2\right)$ - state $\{0\}$, it is also possible to get the state $\{1\}$ with probability $\left(|c|^2 + |d|^2\right)$. If the obtained state is $\{0\}$, then we projected the initial state over the sub-space determined by vectors $|00\rangle, |11\rangle$. The result of this projection is $\left(a'|00\rangle + b'|01\rangle\right)$. Now it is necessary to make a renorming for the second qubit coefficients so that the probability of getting the state $\{0\}$ or $\{1\}$ would be equal to one again in the amount. From the thing that new coefficients $|a'|^2 + |b'|^2 = 1$, it follows that the second qubit state will be:*

$$\frac{a}{\sqrt{|a|^2+|b|^2}}|00\rangle+\frac{b}{\sqrt{|a|^2+|b|^2}}|01\rangle; a'=\frac{a}{\sqrt{|a|^2+|b|^2}}, b'=\frac{b}{\sqrt{|a|^2+|b|^2}}$$

In general, it is a trivial result for two independent qubits; when a d = b c, coefficients may be obtained by multiplication of their states.

Getting back to the properties of entangled states, it is necessary to note that the result of measurement will be a bit different for them. It is clear from the previous one that for a set of independent qubits, the result of measuring the first ones does not give any information about the second. If, for instance, we analyse the state - $\frac{1}{\sqrt{2}}\left(|00\rangle+|11\rangle\right)$, which is entangled, then, in the first measurement, the probability (if we formulate it as «measuring the first qubit», then it implies the presence of two objects but, in fact, there is one system) of getting the state «0» or «1» will be equal to 1/2. However, a 100% coincidence with the first result will be obtained in the second measurement at the same basis, *i.e.* the states of conditionally, the «first» and the «second» qubit turn to be correlated. In other words quantum phenomenon may not be decomposed into independent parts. It is possible to suggest other variants of the entangled pair $\frac{1}{\sqrt{2}}\left(|01\rangle+|10\rangle\right)$ with different signs and coefficients before states $a|00\rangle\pm b|11\rangle$, $a|01\rangle\pm b|10\rangle$. These states differ in the thing that independent qubits may be presented as a tensor product $\frac{1}{\sqrt{2}}\left(|00\rangle+|01\rangle\right)=|0\rangle\otimes\frac{1}{\sqrt{2}}\left(|0\rangle+|1\rangle\right)$, whereas entangled pairs (as they are usually called) may not. In connection with the above-mentioned, there arises a doubt if it is correct to use the term «pair», as it implies the presence of two objects, but not a uniform quantum system. The experiment points out the thing that, in this case, the semantics of this term is not adequate to the situation.

Various physical effects for quantum objects, including their measurement, may be described as the impact of linear operators onto *bra-* and *cket-*vectors that reflect these objects properties.

- *Remember the statement of von Neumann: «Linear operator F, functional in the Hilbert space, corresponds to physical value **F**. Self-conjugated operators correspond to real values, and the corresponding operator functions - to values functions».*

  As vectors are written down as matrices, operators are representable as matrices, respectively. Thus, it is possible to put down that vector $|f\rangle$ or $|f\rangle=\mathbf{A}|s\rangle$ is the result of some linear operator A effect for cket $|s\rangle$, registration $\langle f|\mathbf{A}|s\rangle$ being correspondent to the transition probability of the quantum system from state «s» into state «f». It is used when we pass on from the description of one property to the other or when describing a reversible influence on the quantum system. Then some convenient and known operator's presentation is used in a basis {a} in which decomposition of the initial and finite vectors $|s\rangle$ and $|f\rangle$ is realised; so the full entry will be as follows:

  $$\langle f|\mathbf{A}|s\rangle=\sum_j\sum_i\langle f|\ a_j\rangle\langle a_j|\mathbf{A}|a_i\rangle\langle a_i|\ s\rangle$$

  It is postulated in quantum mechanics that operator A (making up mathematical description) corresponds to each observed physical value, and the mean value of an observed value is determined by the formula:

$$\langle \mathbf{A}\rangle=\int\psi^*(x)A\psi(x)dx=\langle\psi|\mathbf{A}|\psi\rangle,$$
$$\int\psi^*(x)\psi(x)dx=1$$

(3.16)

where $\psi^*\psi$ reflects the density of distribution probability for some variable «x». For the quantity having discrete values, it is possible to write down that:

$$\langle \mathbf{A} \rangle = \langle \psi | \mathbf{A} | \psi \rangle = \sum_j \sum_i \langle \psi | a_j \rangle \langle a_j | \mathbf{A} | a_i \rangle \langle a_i | \psi \rangle = \sum_n |c_n|^2 A_{nn}$$

$$\langle \psi | \psi \rangle = \sum_n |c_n|^2 = 1$$

With the presence of ψ function's norming, the above mentioned statement about the mean value of the observed quantity, expressed by its operator, becomes obvious. It is clear that the value of physical quantity obtained during measuring will always be real.

Generally, the correlations presented by linear equations - where α -an arbitrary complex number - will be true for operators:

$$\mathbf{A}\{|a\rangle + |b\rangle\} = \mathbf{A}|a\rangle + \mathbf{A}|b\rangle$$

$$\mathbf{A}\{\alpha|a\rangle\} = \alpha \mathbf{A}|a\rangle$$

$$\{\mathbf{A} + \mathbf{B}\}|a\rangle = \mathbf{A}|a\rangle + \mathbf{B}|a\rangle$$

According to the rules, the operator affects cket being left of it and right of the bra-vector. However, there is some peculiarity here connected with operators commutativity, which is not realised in our case, i.e. the successive A and B operators effect for any vector may be prsented as $\{AB\}|a\rangle = AB|a\rangle$, the obtained result is being different from vector. It is connected with the thing that linear operators, according to physical interpretation, correspond to dynamic variables - such as co-ordinates, impulse components, etc. It turns out that the measurement's succession at the onset of co-ordinates of the object's impulse has a result different from the case of events reverse sequence. It has already been mentioned above that such events are incompatible in time.

Passing on to the object description is carried out on the basis that corresponds to the property in our focus (devices measuring various characteristics (parameters) are different). These bases, just as properties, are different. They allows us, using *e.g.* Fourier-transformation, to transfer from describing one property to the other.

- *The transfer proper to the object representation is set with the following condition:*

$$\frac{d}{dt} \int_V \psi^* \psi \, dV = 0$$

It means that the probability to detect an object (some properties) in space does not depend on time. Schrödinger equation, which is postulated, is used to find out the energetic spectrum:

$$i\hbar \frac{\partial \psi}{\partial t} = -\frac{\hbar^2}{2m} \frac{\partial^2 \psi}{\partial x^2}$$

In 1926 he simply proposed a kind of equation for wave function. Even though, surely, there are certain analogies with classical physics, e.g. the continuity equation [17, 30]:

$$\int_{\Delta V} \frac{\partial \rho}{\partial t} dV = \oint_{S_0} j(r,t) dS$$

which means that a change in the density of probability ρ in some element of volume ΔV is equal to the flux of probability through the surface of this volume. One can put down the continuity equation passing on to the volume integral and the difference of in- and outcoming flux through divergence as

$$\int_V \left(\frac{\partial \rho}{\partial t} + div\vec{j} \right) dV = 0$$

If it is true for the arbitrary volume element, then it follows from here:

$$\frac{\partial \rho}{\partial t} + div\vec{j} = 0$$

As it is possible to present the density of probability as $\int_V \Psi\Psi^* dV$, then the temporal fluxion will be:

$$\frac{\partial}{\partial t} \int_V \Psi\Psi^* dV = \int_V \left(\Psi^* \frac{\partial \Psi}{\partial t} + \Psi \frac{\partial \Psi^*}{\partial t} \right) dV$$

One can represent the determination of probability flow as $\vec{j}(r,t) = -C \cdot grad\rho$. Herewith:

$$div\vec{j} = div\left(-C \cdot grad\left(\int_V \Psi^*\Psi dV \right) \right)$$

and for one-dimensional case this equation is confined to:

$$-C\int_V div\left(grad\left(\Psi^*\Psi \right) \right) dV \Rightarrow -C\int_x \left(\Psi^* \frac{\partial^2 \Psi}{\partial x^2} + \Psi \frac{\partial^2 \Psi^*}{\partial x^2} \right)$$

Summing it with the temporal fluxion, we will finally get the following equation:

$$\int_x \left\{ \left(\Psi^* \frac{\partial \Psi}{\partial t} + \Psi \frac{\partial \Psi^*}{\partial t} \right) - C\left(\Psi^* \frac{\partial^2 \Psi}{\partial x^2} + \Psi \frac{\partial^2 \Psi^*}{\partial x^2} \right) \right\} dx = 0$$

and it follows from here that, under the integral, the equation is to turn to zero if it is true for any «x». It is not difficult to note that this expression is the addition of two equations like:

$$\frac{\partial \psi}{\partial t} = C \frac{\partial^2 \psi}{\partial x^2}$$

for the probability amplitude and the value complexly conjugated with it. Thus, Schrödinger equation expresses the continuity of the density of an object's detection probability in the transition from one representation to the other.

On the other hand, we can considerably mix up the situation using analogies with classical physics. Thus, analysing some phenomena limited by spatial interval «d» means pointing out a quasi-closed system. If it is possible to consider some phenomena, then they are only physical in it. Fourier transformation of single impulse (Fig. (3.20), whose zeroes result in modes ω_k that correspond to a particle's impulse representation in a potential well, is well known in quantum mechanics.

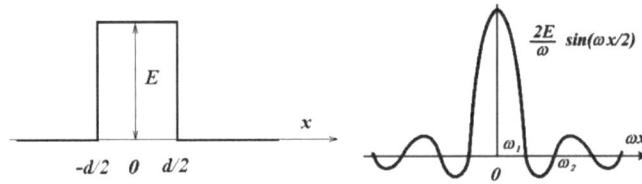

Figure 3.20. Fourier transformation of single impulse.

$$S(\omega) = \int_{-\frac{d}{2}}^{\frac{d}{2}} E e^{-j\omega x} dx = \frac{E}{-j\omega} e^{-j\omega x} \Big|_{-\frac{d}{2}}^{\frac{d}{2}} =$$

$$= \frac{2E}{\omega} \sin \frac{\omega x}{2} \Rightarrow \lambda_n = \frac{d}{n} \; (n = 1, 2 \ldots) \Rightarrow \omega_k = \frac{2\pi k}{d}$$

The presence of still water means a whole number of half-periods of certain oscillations frequency to be within the spatial interface. If we consider the thing within the accepted conditionality, then the wave function - to which this wave is ascribed - reflects the amplitude of probability to detect a particle within the potential well. However, it is impossible to check this interpretation experimentally without changing the properties of the initial system. To distribute ω_k over energy, it is necessary to introduce rigidness in the system's description. For this purpose, the Schrödinger equation is used in quantum mechanics on the analogy with classical physics. But de Broglie ratios for «quasi»-impulse and energy point out a principally different type of presentation, not the object one.

WAYS OF MEASUREMENT PROCESS DESCRIPTION

Von Neumann projectors, collapse of wave function.

Measurement in quantum mechanics is a separate huge problem which may be only outlined here. First, we mention the commonly accepted description which leads to the known paradox of wave function's collapse. Following the material formulated in the book [31], let us consider the operators which realise the quantum system's projection with wave function $|\psi\rangle$ into one of the basis states $|a\rangle$. Such operators are called *von Neumann projectors*, as Neumann is the first who made such a description.

Let it be possible to get some of the states $|a_1\rangle$ during the measurement of quantum system $|\psi\rangle = c_1 |a_1\rangle + c_2 |a_2\rangle$ with a certain probability $p_1 = |c_1|^2$, or $|a_2\rangle$ (with a probability $p_2 = |c_2|^2$) *i.e.*

$$|\psi\rangle = c_1 |a_1\rangle + c_2 |a_2\rangle \Rightarrow \begin{cases} |a_1\rangle, & p_1 = |c_1|^2 \\ |a_2\rangle, & p_2 = |c_2|^2 \end{cases}$$

which may be written as operator-projectors:

$$\begin{cases} \mathbf{P}_1 |a_1\rangle = |a_1\rangle, \mathbf{P}_1 |a_2\rangle = 0 \\ \mathbf{P}_2 |a_2\rangle = |a_2\rangle, \mathbf{P}_2 |a_1\rangle = 0 \end{cases}$$

Each operator $\mathbf{P}_i$ transfers the initial state to only one basis a_i, and the interaction of quantum system being, in a general case, in the state of superposition, occurs with the measurement device and the surrounding. As a result, a confusion in the measured system's state and the measurement device takes place. Sometimes, in a big device, a microsensor that can «feel» the quantum system is pointed out, and the change of its state may be interpreted as a record of information about the measured system. Such a microsensor may be called a grain in the photofilm sensitive layer, photodetector, Wilson chamber, *etc.* If we successively stick to the points of quantum mechanics, then there is no reversibility at the physical level. One can think that the system and the measurement device with the surrounding turn in the state of superposition. Such interpretation leads to multi-dimensional *Everett Universes* [14, 15], but, finally, it does not solve the problem of the initial wave function's collapse.

In the standpoint of quantum mechanics, both the device and the observer are also quantum systems. Let the initial state of an object be $|\psi_0\rangle = c_1|a_1\rangle + c_2|a_2\rangle$, device ϕ_0 and observer φ_0. We indicate the initial state of the whole system as $|\psi_0\rangle|\phi_0\rangle|\varphi_0\rangle$. After the measurement, the system of device-observer will transfer to state ϕ_1 and φ_1, which are indicative of the thing that the device changed its values and the observer perceived it. Hence, the whole system changed its state: $|\psi_0\rangle|\phi_0\rangle|\varphi_0\rangle \rightarrow |\psi_1\rangle|\phi_1\rangle|\varphi_1\rangle$ - and that corresponds to the case of basis «1» perception of the observed state. Analogously, for the second possibility of an object's interpretation, we can write it down as $|\psi_0\rangle|\phi_0\rangle|\varphi_0\rangle \rightarrow |\psi_2\rangle|\phi_2\rangle|\varphi_2\rangle$, and that corresponds to the case of basis state «2» perception of the observed system.

According to the rules of quantum mechanics, we can write the situation of the measurement as follows:

$$\left(c_1|\psi_1\rangle + c_2|\psi_2\rangle\right)|\phi_0\rangle|\varphi_0\rangle \rightarrow \left[c_1|\psi_1\rangle|\phi_1\rangle|\varphi_1\rangle + c_2|\psi_2\rangle|\phi_2\rangle|\varphi_2\rangle\right]$$

(the so-called entangled state of the measurement system or the surrounding); after the act of measurement, using von Neumann operator-projectors, we will obtain one of two possible finite states:

$$\mathbf{P}\left[c_1|\psi_1\rangle|\phi_0\rangle|\varphi_0\rangle + c_2|\psi_2\rangle|\phi_0\rangle|\varphi_0\rangle\right] \Rightarrow \begin{bmatrix} c_1|\psi_1\rangle|\phi_0\rangle|\varphi_0\rangle \\ c_2|\psi_2\rangle|\phi_0\rangle|\varphi_0\rangle \end{bmatrix}$$

It means that the measured «object» was in the state of superposition before the measurement $\left(b_1|\psi_1\rangle + b_2|\psi_2\rangle\right)$, the device and the observer being in the initial state $|\phi_0\rangle|\varphi_0\rangle$; after the measurement, the full system will be in the state of superposition $c_1|\psi_1\rangle|\phi_1\rangle|\varphi_1\rangle + c_2|\psi_2\rangle|\phi_2\rangle|\varphi_2\rangle$ till some «moment of choice» after which only one measured value remains and the collapse of wave function occurs.

In his contribution [12], von Neumann focussed on the problem in detail: whether it is possible to draw the line between the described reality and the perception? but the answer is no. There is no such line and this is the only way out of this situation. We are always within our feelings, and only the way of description determined by a situation is changing.

- *Let us consider an example that illustrates the measurement process of a quantum system under the interaction with a classical device [32, 33]. Taking interest in the direction of mass μ microparticle - to make it simple - having one degree of freedom. The measurement proceeds with the interaction of the macroparticle - a ball of mass M being in an unstable position on the top with a very small deepening. Before the interaction, the microparticle is described by wave function $\psi_0(\xi) = A^+ e^{ik\xi} + A^- e^{-ik\xi}$, where ξ is microparticle's co-ordinate, k - its impulse, and microsystem of mass M can be presented by wave function*

$$\Psi_0(Q) = \frac{1}{\sqrt{\pi}} e^{-\frac{Q^2}{2a^2}}, \text{ where } a = \sqrt{\hbar/M\omega_0}, \quad \omega_0 \text{ - ball fluctuation frequency, } Q \text{ - its co-}$$

   ordinate. Before the collision, at moment t_0, the whole system's wave function will be:

$$\Phi(Q,\xi,0) = \Phi(Q,\xi) = \Psi_0(Q)\psi_0(\xi),$$

where Q - ball's co-ordinate, ξ - particle's co-ordinate. The hamiltonean of Schrödinger equation describing this system is set by the expression:

$$H(Q,\xi) = -\frac{\hbar^2}{2M}\frac{\partial^2}{\partial Q^2} + U(Q) - \frac{\hbar^2}{2\mu}\frac{\partial^2}{\partial \xi^2} + W(Q,\xi)$$

Microparticle μ is thought to be free; ball M is in the potential field $U(Q)$. Value $W(Q,\xi)$ is the particle's energy (which is supposed to be point) of interaction with the ball: $W(Q,\xi) = g\delta(Q-\xi)$. The solution of Schrödinger equation is searched as:

$$\Phi(Q,\xi,t) = \Phi_0(Q,\xi) + \Phi^+(Q,\xi,t) + \Phi^-(Q,\xi,t)$$

in the first approximation of perturbation theory. As a result of calculations, it turns out that function $\Phi^+(Q,\xi,t)$ is different from zero at $t \to \infty$ only in domain $0 < Q < \infty$, which corresponds to the ball's falling on the right (positive value of the microparticle's impulse). In the second case $\Phi^-(Q,\xi,t)$, the ball falls down from the well on the left at values $-\infty < Q < 0$. Probability density $\rho(Q,\xi,t,Q',\xi',t) = \Phi^*(Q,\xi,t)\Phi(Q',\xi',t)$ is put down in general using 9 summands:

$$\begin{aligned}
\rho(Q,\xi,t;Q',\xi',t) = {} & \Phi_0^*(Q,\xi,t)\Phi_0(Q',\xi',t) + \\
& + \Phi_0^*(Q,\xi,t)\Phi^+(Q',\xi',t) + \Phi_0^*(Q,\xi,t)\Phi^-(Q',\xi',t) + \\
& + \Phi^{+*}(Q,\xi,t)\Phi_0(Q',\xi',t) + \Phi^{-*}(Q,\xi,t)\Phi_0(Q',\xi',t) + \\
& + \Phi^{+*}(Q,\xi,t)\Phi^-(Q',\xi',t) + \Phi^{-*}(Q,\xi,t)\Phi^+(Q',\xi',t) + \\
& + \Phi^{+*}(Q,\xi,t)\Phi^+(Q',\xi',t) + \Phi^{-*}(Q,\xi,t)\Phi^-(Q',\xi',t)
\end{aligned}$$

At ($t \to \infty$) and $|Q|, |Q'| > a$, all the members of this matrix disappear with the exception the last two. Functions Φ_0 disappear at $Q, (Q') \to \pm\infty$ as the functions describing the ball's position in the pit as $e^{-Q^2/2a^2}$ also disappears. Calculations show that Φ^- disappears at $Q' \to +\infty$: it describes the situation in which the balls turn to be left of the pit, i.e. at negative values of Q', and vice versa Φ^+ disappears at $Q' \to -\infty$.

Thus, using the quantomechanical description, it is possible to get the observation probability of a classical result - ball's falling onto the right or left, whith a certain direction of the particle's impulse. Herein, one should focus on in reality, observation lasts for a finite time, and our spatial-temporal description coincides with a possible result only after the integration over the whole interval from $-\infty$ to $+\infty$. Replacement of one time by the other, and the transfer of notions from «space of description» into reality leads to the paradox of wave function's reduction.

TWO-LEVEL QUANTUM SYSTEM

Ammonia molecule, Hamilton operator, state evolution in time for a system's arbitrary energy.

The quantum system's states described by its properties (impulses, moments of movement quantity, spins, *etc.*) are calculated using the Schrödinger equation. Distribution of a two-level system's states probability is of special focus, as it may be used as qubit when making up a quantum computer register. Ammonia molecule (NH_3) can be a two-level system if we are only interested in the position of nitrogen atom molecule related to the plane hydrogen atoms lie in [16].

- *This molecule is a regular pyramid formed by hydrogen atoms at its base and nitrogen atoms above them. One of the interesting perculiarities of NH_3 molecule is that the nitrogen atom may have two stable states - over the plane where hydrogen atoms lie in and under this plane. It turns out that the presence of spin in atoms nuclei forming a molecule and their interaction in electric and magnetic fields leads to inverse segregation of energy levels. Thus, the nitrogen position in a molecule related to the plane where hydrogen atoms lie in is a physically differentiated state. This difference is small in energy, i.e. about 10^{-4} eV. However, it is on these transitions that the very first laser was developed with radiation wavelength ≈ 1.3 cm, which is equal to radiation frequency 24 GHz.*

First, let us consider some properties of Hamilton operator which is used for the calculation of quantum system's states. Suppose there is some state $|\psi(t)\rangle$, where t - some parameter, and its decomposition on the chosen basis $\sum_n |i\rangle$ will be $c_i(t) = \langle i | \psi(t)\rangle$. We can think this parameter to be a system's «internal time» which changes from plus (positive) to minus (negative) infinity. Then the dependence of wave function on this parameter will be:

$$-i\hbar \frac{dc_i(t)}{dt} = \sum_j H_{ij} c_i(t)$$

where the sum is considered on the corresponding Hamilton matrix (energy operator) parameters. As full probability of a system's detection is normed to 1 and does not depend on time, then: $\frac{d}{dt}\sum_i c_i(t) c_i^*(t) = 0$.

It follows from here that (after substitution of the previous formula and summands regrouping when summing up on both indices):

$$\sum_i \sum_j \left(H_{ij} - H_{ji}^*\right) \cdot c_j c_i^* = 0 \Rightarrow H_{ij} = H_{ji}^*$$

It means that the matrix elements symmetrically positioned in relation with the main diagonal are complex conjugated values, and the diagonal elements will have the real values [16].

- *Actually, the object quantum system's description is presented using classical analogies. It is for this reason that it looks unnatural. It has already been mentioned above that, under Fourier transformation, the superposition of spatial limits for the system allows us to get a set of harmonics connected with some stationary states. The transition from one level to the other corresponds to certain energy. We are now focussed on the analysis of such transitions possibility when describing the quantum system. If there was a way of rigidness introduction, e.g. using Fourier «window» transformation, then the Schrödinger equation would be excessive.*

For a two-level system, in general, the Hamilton matrix will be as follows:

$$i\hbar \frac{dc_1(t)}{dt} = H_{11} c_1 + H_{12} c_2$$

$$i\hbar \frac{dc_2(t)}{dt} = H_{21} c_1 + H_{22} c_2$$

(3.17)

If we set all H_{ij} equal to zero, we will get the system of equations like:

$$i\hbar\frac{dc_1(t)}{dt}=H_{11}c_1$$
$$i\hbar\frac{dc_2(t)}{dt}=H_{22}c_2$$

(3.18)

And its solutions will be as follows:

$$c_1=(const)\exp\left(-\frac{i}{\hbar}H_{11}\,t\right),\quad c_2=(const)\exp\left(-\frac{i}{\hbar}H_{22}\,t\right)$$

(3.19)

It means that in the measurement, it is possible to obtain the energy values of either H_{11} or H_{22} (which is adequate to nitrogen atom's location in one of the extreme positions), and these states may be set as basis.

Now we will observe the change with the occurrence of non-diagonal matrix elements. It corresponds to the case when the system has some energy E which differs from basis values. But the question is how is it possible with discrete energy levels? We have to accept it to explain the case of «superpositional» state because external effects will change the system's state in a way. Now have to find the decomposition of state $|\psi(t)\rangle$ on basis vectors $|\psi\rangle=|1\rangle\langle1|\,\psi\rangle+|2\rangle\langle2|\,\psi\rangle$, *i.e.* it may be represented as a sum of basis vectors with some coefficients that correspond to projections onto these orths.

Let us indicate coefficient $\langle1|\,\psi\rangle$ as C_1 and $C_2=\langle2|\,\psi\rangle$. We then need to solve equation (3.18) considering the properties of non-diagonal matrix elements to be designated as A. This value reflects a change of the system's detection probability in the first or second state when measured:

$$i\hbar\frac{dC_1(t)}{dt}=EC_1-AC_2$$
$$i\hbar\frac{dC_2(t)}{dt}=-AC_1+EC_2$$

(3.20)

These equations may be solved as follows [16] - let us add the first to the second:

$$i\hbar\frac{d}{dt}(C_1+C_2)=(E-A)(C_1+C_2)$$

and we will obtain the solution like:

$$C_1+C_2=ae^{-\frac{i}{\hbar}(E-A)\cdot t}$$

(3.21)

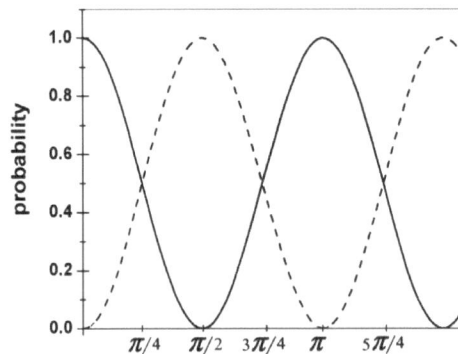

Figure 3.21. The probabilities of molecule's detection in state «1» and «2».

Then, subtracting one from the other will result in:

$$i\hbar \frac{d}{dt}(C_1 - C_2) = (E + A)(C_1 - C_2)$$

and it will lead to:

$$C_1 - C_2 = b e^{-\frac{i}{\hbar}(E+A)\cdot t} \tag{3.22}$$

It is necessary to choose the constants of a and b integration so that there would be matching initial conditions for the present physical task. Adding and subtracting (3.21) and (3.22) will result in C_1 and C_2:

$$C_1(t) = \frac{a}{2} e^{-\frac{i}{\hbar}(E-A)\cdot t} + \frac{b}{2} e^{-\frac{i}{\hbar}(E+A)\cdot t} \tag{3.23}$$

$$C_2(t) = \frac{a}{2} e^{-\frac{i}{\hbar}(E-A)\cdot t} - \frac{b}{2} e^{-\frac{i}{\hbar}(E+A)\cdot t} \tag{3.24}$$

Now we are going to choose (at $t = 0$), the initial conditions - those adequate to the molecule's position in state «1», *i.e.* $C_1(0) = 1$ and $C_2(0) = 0$. It corresponds to equations: $C_1(0) = \frac{a+b}{2} = 1$, $C_2(0) = \frac{a-b}{2} = 0$ from where it follows that $a = b = 1$. Using them for formulas in $C_1(t)$ and $C_2(t)$ and making the common multiple, we will get:

$$C_1(t) = e^{-\frac{i}{\hbar}E\cdot t} \left(\frac{e^{\frac{i}{\hbar}A\cdot t} + e^{-\frac{i}{\hbar}A\cdot t}}{2} \right)$$

$$C_2(t) = e^{-\frac{i}{\hbar}E\cdot t} \left(\frac{e^{\frac{i}{\hbar}A\cdot t} - e^{-\frac{i}{\hbar}A\cdot t}}{2} \right)$$

Whence, according to the Eiler formula for sine and cosine, it follows that:

$$C_1(t) = e^{-\frac{i}{\hbar}E\cdot t} \cos\frac{A\cdot t}{\hbar} \tag{3.25}$$

$$C_2(t) = i\cdot e^{-\frac{i}{\hbar}E\cdot t} \sin\frac{A\cdot t}{\hbar} \tag{3.26}$$

The value of both amplitudes is harmonically changing with «time», and the probabilities of the thing that the molecule will be found in state $|1\rangle$ or $|2\rangle$, at a certain parameter t value, will be equal to squared $C_1(t)$ and $C_2(t)$ modules, respectively:

$$\left|C_1(t)\right|^2 = \cos^2\frac{A\cdot t}{\hbar} \tag{3.27}$$

$$\left|C_2(t)\right|^2 = \sin^2\frac{A\cdot t}{\hbar} \tag{3.28}$$

In Fig. (**3.21**), you can see that the probabilities of molecule's detection in state $|1\rangle$ and $|2\rangle$ concertedly change depending on parameter $\dfrac{A}{\hbar}t$ being totally equal to one. Thus, the nitrogen atom can slip past among hydrogen atoms (tunnely leak under the barrier) and be detected in this or other basis state. It is also clear how it is possible to change the quantum system's state reversibly. It can be achieved by radiation impulse effect of certain duration with the energy equal to that of transition between levels.

- *The terms «space» and «time» were used in this and the previous paragraphs and are believed to be intuitively clear. At the same time, these concepts have been known for millenia and they will be discussed in more detail further on.*

CONSCIOUSNESS, SPACE AND TIME

Psychological and physical times; problem of determining past, present, and future; determinism and evolution; limitations of finite theories in description of reality; consciousness and reality.

Many relational aspects of space, time, consciousness and the paradoxes connected with it, which are now widely discussed, were first most acutely raised by Aurelius Augustine in his «Confession». Consciousness is our Reality. There is no us without consciousness - no memory, no thinking. If our consciousness reflects outer reality, then it exists in it somehow - let it be not in space, but, surely, it is in time. How can consciousness exist in the «physical» time? Where does the time come from and where does it go to, and where is Consciousness? This enigma is more than two thousand years old, but there is still no answer of it.

We can remember St. Augustine's frenetic questioning, his prayer to God [34]:

I ask, Father, I do not affirm. O my God, rule and guide me. Who is there who can say to me that there are no three times (as we learned when boys, and as we have taught boys) the past, present and future, but only present, because these two are not? Or are they also; but when from future it becometh present, cometh it forth from some secret place, and when from the present it becometh past, doth it retire into anything secret? For where have they, who have foretold future things, seen these things, if as yet they are not? For that which is not cannot be seen. And they who relate things could not relate them as true, did they not perceive them in their mind. Which things, if they were not, they could in no wise be discerned. There are things both future and past... I have just now said, then, that we measure times as they pass, that we may be able to say that this time is twice as much as that one, or that this is only as much as that, and so of any other of the parts of time which we are able to tell by measuring. Wherefore, as I said, we measure times as they pass. And if anyone should ask me, «Whence dost you know?» I can answer, «I know, because we measure; nor can we measure things that are not; and things past and future are not.» But how do we measure present time, since it hath not space? It is measured while it passeth; but when it shall have passed, it is not measured; for there will not be aught that can be measured. But whence, in what way, and wither doth it pass while it is being measured? Whence, but from the future? Which way, save through the present? Whither, but into the past? From that, therefore, which as yet is not, through that which hath no space, into that which now is not. But what do we measure...?

We can only be astonished at St. Augustine's acumen who asked the most critical questions about the nature of time which also remain topical now. What is the past, present and future? If they are qualitatively different, then how can we place them on one temporal scale? If it is a «phase transition» like transformation of water into ice or vapour, then what scale is it on? Does the time flow? Has it any direction (gradient)? Has it any beginning or end? One of the most profound enigmas is the one which Sir Arthur Eddington called «time arrow».

We naturally accept the things when events continue in time in a certain direction: people grow old, pots are broken, candles burn out. If there were no temporal asymmetry, then people would grow younger as often as they would do older. But, in the laws of physics, time is completely symmetrical as a parameter, and it does not «flow» or «fly» anywhere. Physically, there is no difference between the past, present and future. Elementary particles may come from both the future and the past - it is set in the equations of quantum dynamics, and it is experimentally proved with unsurpassed precision.

In Einstein's General Theory of Relativity (GTR), *Universum* is described as a unified object, as something granted in the space-time. But it is for God that everything is granted. The attempt to describe *Universum* from the «by-side» has brought forth a number of problems. One of them is a cosmological paradox. It follows from GTR that our Universe is not stationary, and cosmology ascribes it the age of about 15 milliard years. It is obvious that the «Great Explosion», which was the genesis of the Universe, is an event, but events do not belong to the traditional formulation of the laws of nature. Trajectories and wave functions neither begin nor end. And, herewith, we cannot help remembering St. Augustine [34]:

> For in what temporal medium could the unnumbered ages that thou didst not make pass by, since thou art the Author and Creator of all the ages? Or what periods of time would those be that were not made by thee? Or how could they have already passed away if they had not already been? Since, therefore, thou art the Creator of all times, if there was any time before thou madest heaven and earth, why is it said that thou wast abstaining from working? For thou madest that very time itself, and periods could not pass by before thou madest the whole temporal procession. But if there was no time before heaven and earth, how, then, can it be asked, «What was thou doing then?» For there was no «then» when there was no time.

The medievel question about the thing «what was God doing before the Creation of the world?» became topical again for physicists. The occurrence of Being out of Nothing is a creative act. But creativity contradicts to the physical interpretation of time. If we can describe changes of the form in finite theories with a limited and unchanged content, then how is it possible to describe the changes of content proper? It is impossible to develop a classical theory whose axiomatics is not determined or spontaneously changing. Yet, Platon connected reason and truth with the access to Being - unchanging Reality behind Becoming. However, he also realised the contradictious character of such a standpoint, and in his «Sophist», he came to the conclusion that we need both Being and Becoming. The same problem was come across by antique atomists. To permit the appearance of a new content, Lucretius Carus [35] had to introduce «klinamen» that excites atoms deterministic motion - a sort of some ratio of indefiniteness:

> «We wish thee also well aware of this:
>
> The atoms, as their own weight bears them down
>
> Plumb through the void, at scares determined times,
>
> In scares determined places, from their course
>
> Decline a little - call it, so to speak,
>
> Mere changed trend. For were it not their wont...»

In 2.5 thousand years we come across the analogous statement in the Einstein's contribution devoted to spontaneous light emission by an excited atom [36], where it goes about the thing that time and direction of elementary processes are randomly determined. It is inevitable as creativity does not belong to physics, and creativity refers to physicists, but not their theories. The time arrow could be correlated with the direction of evolution as a process of content changing, and creativity belongs not to the object world but to the subject (apropos, going about Evolution in this sense, researchers - as if they would agree upon it - write the word Nature with a capital letter and indirectly imply the subject in it). For Ludwig Baltzman, a

Viennesse physicist, the introduction of time in physics as a notion connected with evolution was the reason for his whole life. Earlier, klinamen and creativity were denied by science as phenomena that lead to the aberration of causality in the world. However, the latest experiments (A. Aspect *et al.*, [37, 38]) in studying the so-called EPR effects [45], connected with states quantum teleportation, are vividly indicative of the thing that Nature does not prohibit such processes.

Prediction (prognosis) is the most important aim of the scientific theory. But prediction is the future and it has always been thought of as ideal, but not real. It could not affect the present or past. However, the phenomena of interference of a quantum particle's wave functions in this very «future» explicitly determine its present and past. Due to this, e.g an electron can simultaneously transmit through two slots if it is not «peeped». Hence, the future is not a mental construction, but objective reality being, in a form, currently present. But what is present? If there were no our mind that always «experiences» this present, then who would indicate this point in time? And is it a point? That is what St. Augustine [34] wrote about it:

> But even now it is manifest and clear that there are neither times future nor times past. Thus it is not properly said that there are three times: a time present of things past; a time present of things present; and a time present of things future. For these three do coexist somehow in the soul, for otherwise I could not see them. The time present of things past is memory; the time present of things present is direct experience; the time present of things future is expectation. If we are allowed to speak of these things so, I see three times, and I grant that there are three. Let it still be said, then, as our misapplied custom has it: «There are three times: past present, and future.» I shall not be troubled by it, nor argue, nor object - always provided that what is said is understood, so neither the future nor the past is said to exist now. There are but few things about which we speak properly - and many more about which we speak improperly - though we understand one another's meaning.

True, it is inevitable that,when considering time, we introduce some concepts - rather psychological than physical. For instance, the point where those very laws of nature, which an electron observes in its motion, are recorded? There is no way they can be recorded in the particle proper, nor it can determine its trajectory of motion. These laws belong to all *Universum* as a whole. But then we speak about the laws of Nature *per se*, but not the separate objects. If we consider the time representaton in our consciousness, it turns to be dual. On the one hand, we experience the time as some current in which our consciousness is always in the «present» and is irreversibly involved from the past into the future. In this thing it differs from the «static» space in which all dots are available to us. But, on the other hand, we can «travel» in our past making it to be present for a while with the help of memory, or we can travel in the future using our imagination. This aspect of temporal perception is like that of spatial one. How does «future-in-the-present» of a subject correlate with the physical «present-of-the-future» of an object? Experiments in hypnotic regression really return our organism, physiologically, to childhood. There even appears an infantile reflex - nostrils collapse - which adults do not have. Besides, there are numerous documentary artefacts registered in history when people really forefelt the future. Any creative act is a transfer from Nothing into Being, and it is an aberration of causality. When Mendeleev discovered his periodicity table, there was yet no atomic physics that explains the cause of its existence. Then, what does the subjective future differ in from the objective future of quantum world?

When a painter creates a picture or a conversation goes on, all the unfinished process refers to the present. Does time exist as a moment? St. Augustine denied it. True, to perceive a sonic tone, it is necessary to determine its frequency physically. But to determine it physically there should be a span of time equal minimum to the period of one fluctuation (sonic). And to set one harmonic physically, the whole spatial-temporal continuum is needed. Does the time flow anywhere else beyond our consciousness? Neurophysiological investigations show us that retina photoreceptors integrate into receptive fields and project themselves onto outer geniculate body, where they begin to pulsate periodically collapsing into one point. During this process, we go «blind» for a while.

---

[45] originating from the known paradox formulated by Einstein, Podolsky and Rosenov.

Thus, perception continuity in our consciousness is an illusion made by specific brain sections. In particular, people with stroke-affected V5 field see the outer reality as a sequence of immobile still-frames. Crossing a street, they run a big risk, as they perceive moving vehicles as immobile objects which jerkily change their spatial position. Actually, we perceive the world on «frames», and perception continuity is an illusion. Probably, is Einstein Universe just a frame in the perception of Nature from one creative act to the other? In his contribution of 1912 «Quantum hypothesis», in which the points of relativity theory were not considered at all, the statement about discreteness of a multitude of possible states for any isolated physical system - as Poincare mentions - is also applicable to the Universe: «Hence, the Universe is to jerkily transform from one state to the other, but it remains unchanged in the intervals among «jumps». And different moments during which it preserves its state would not be distinguishable from each other; thus, we approach to the breakable time flow, - to the atoms of time.»

Our common consciousness represents time as something regularly (evenly) flowing from the «past» into the «future». To represent such a position, it is necessary to introduce some «supertime» t' on which it would be, at least, possible to determine the velocity of «time flow» to make sure it is constant. Besides, we need to be sure that this other time «flows evenly» and so on till the infinity. However, if there is some pause of it for a period of $\Delta t'$, we will not be able to notice it, as all spatial process will be just «frozen» during this time $\Delta t'$. Obviously, the introduction of time t' for time t is senseless. Only interval evaluation of time Δt is important and available for us. It is this that changes during the transfer from one physical countdown system to the other. The paradox consists in that the countdown of system proper is determined by a subject-observer and one's consciousness. It means that, objectively, Lorentz ratios in special relativity theory determine changes in a subject's spatial and temporal perception, but not the space and time proper. In general relativity theory, temporal and spatial perceptions depend on gravitation and are determined by a compression and extension degree of the existing four-dimensional space-time. But is it possible to create a non-contradictious reality concept having dropped its second half off board - the subject and consciousness? The quantor «exists» itself is just a fixation of the fact that something is available, directly or indirectly, to consciousness and perception. Moreover, our consciousness does not need any confirmation of its existence, unlike the «presentive world» on the other side of perception. In a sense, this is the Platonean world of ideas beyond and besides the subject requiring an act of belief for its recognition, as the psycho-physical problem has no solution. So, still, what «compresses and extends» in the «on-the-other-side» world? This is what St Augustine says: *«But no certain measure of time is obtained this way. From this it appears to me that time is nothing other than extendedness; but extendedness of what I do not know. This is a marvel to me. The extendedness may be of the mind itself. For what is it I measure, I ask thee, O my God. «This time is longer than that,» or, more precisely, «This is twice as long as that»* [34].

Another possibility of determining the time arrow arises during the so-called reduction of wave function which occurs at the moment of an object's perception. The latter consequence really made physics to be on the edge of solipsism and caused heated debates. Werner Heisenberg, one of the founders of new physics, turned to be involved in these philosophical and humanistic problems. In «Philosophical problems of quantum physics» he wrote that physicists should neglect the ideas about the objective temporal scale, common for all observers, and that of temporal and spatial events independent from our ability to observe them. Heisenberg emphasised that instead of elementary particles, the laws of nature now deal with our knowledge about these particles, *i.e.* with the content of our reason. In 1958, Erwin Schrödinger, who formulated the basic equation of quantum mechanics, wrote a short book entitled «Reason and matter». In this series of essays, from the results of new physics, he came to a mystical opinion about the Universe, which he identified with «eternal philosophy» of Aldous Huxley. Schrödinger is the first among theoreticians of quantum physics to have expressed his sympathy to the ideas of «Upanishad» - and oriental philosophical thought. Scientific views of many modern physicists are summed up in the essay «Notes on the problem of Reason-Body» written by Eugene Wigner - a Nobel Laureate. At the beginning he mentions the fact that majority of the physicists returned to the recognition of the thing that thought is primary. He asserts: «One could not formulate non-contradictious laws of quantum mechanics by not including consciousness in them». And, in conclusion, he points out the thing that how wonderful it is that scientific studies of the world led us to the content of our consciousness as to the primary Reality.

In the review of his own papers, in 1901, Poincare writes that: «I was focussed on the analysis of psychological initial outsets of the concept of space». The psychological aspect of concepts origin not simply arose his interest, but it was a relevant element of his methodological setting, being spread beyond geometrical notions. Thus, for instance, when connecting his refusal from the notion of actual infinity with psychological speculations, he writes:»Russel will, undoubtedly, tell me that he deals not only with psychology, but also with logic and cognition theory; I will have to answer that there are no logic and cognition theory being independent from psychology [39]«. The most important of this analysis is the conclusion determining influence of our representations to the outer world perception and its geometry. Poincare begins the chapter «Space and geometry» in the book «Science and hypothesis»(1902), as he writes, with a «little paradox» that consists in the creatures whose mind and organs of feelings would be like ours «could get their impressions from the correspondingly matched outer world» so that they would have to build up the geometry, other than that of Euclid and to place the phenomena of this outer world even in the four-dimensional space. In Poincare's contribution of 1906, which is independent from that known of Einstein (1905), there is practically the whole STR apparatus. Mathematically, it seems that this contribution also overlaps that of Einstein and those of Minkowski (1907-1908), with whom the introduction of four-dimensional space-time in physics is connected - the so-called Minkowski space. Supporters of Poincare's priority in the discovery of four-dimensional space-time allocate Minkowski just greater flatness and pathos in comparison with Poincare.

Everything available to us is available only in consciousness and through perception. And, if there is another reality beyond consiousness, then all what we have is only its reflection or a mental map after V.F. Petrenko [26]. As Gregory Bateson writes: «any descriptions, explanations or representations, with the necessity, are the reflection of differential characteristics of a described phenomenon onto some surface, matrix or co-ordinate system. In case of a geographical map, a flat finite-sized sheet of paper is the perceiving matrix, and there are difficulties if the reflected is too big or, *e.g.*, spherical-shaped. Other difficulties would arise if a tor (bagel) surface or a discrete sequence of dots would be the perceiving matrix. Any perceiving matrix, including language or a tautological net of theorems, has its formal characteristics which will principally distort a phenomenon reflected onto it. Probably, our world was projected by Procrustean - not a kind personage from Greek mythology in whose inn all guests were levelled with the length of bed by cutting off and stretching legs [40]«. It means that all we have for making up our theories of «outer» Reality is a mental map - its reflection in some semantic «sub-space» of our consciousness. But «map is no theory, and name is in no way a named thing», as a famous principle of Alfred Korzhibsky reads. Not only colour paints, sounds, *etc.* are absent in the world of scientific matter, but also space which we perceive with vision or touching. It is relevant for science that its matter is in space, but this very space cannot be the same one we see and touch...real shape science is focussed on real space which should be different from the one that seems to every man [41]».

«Then what do we measure?», questioned blessed Augustine. In the philosophical papers of Henri Bergson, time or «duration» plays the major role when discussing the limits of science and the interrelation of man and nature. According to Bergson, our understanding of nature should be based not on the objects outlined by science as a result of their temporal behaviour, but on our own subjective experience, which is, first, is the experience of duration (longevity) and creativity. One of the aims Bergson pursued when working at his «Creative evolution» was the intention to show that «the Whole has the same nature as I, and we comprehend the whole by more and more profound comprehension of I [42]». Does it mean that there is nothing but Consciousness proper? But we admit that any «individual consciousness» is limited and this limit-border may change depending on time and circumstances confining, at times, only to the realisation of «I am». And it means that «I» is only a limited part of «impersonal» Consciousness.

In physics, there is a known paradox of «Schrödinger's cat» in which it is presented in an isolated box. Some radiation substance and an *x*-ray sensor are put into the box. When spontaneous nuclear segregation occurs, the sensor reacts and sets the device that breaks the bulb of cyanide. According to quantum mechanics, before opening the box, the cat was in a virtual state of in-between life and death, and it is only the moment of a subject's perception that determines the choice of real state. The paradox of Wigner's friend is a complicated experiment of Erwin Schrödinger. Eugene Wigner introduced the category of friends. After the end of the experiment, the experimentator opens the box and sees the cat alive. The state vector of the cat, at the moment of opening the box, transfers to the state «nucleus did not decay, cat - alive». Thus, the cat was found alive in the laboratory. The friend was outside the lab. He does not know yet if the cat is alive or dead. The friend will recognize the cat to be alive as soon as the experimentator tells him the outcome of this experiment. But the rest of the friends have not yet admitted the fact that the cat is alive, and they will do so as soon as they are told the experimental result. Thus, the cat may be thought alive only when all the people of the Universe come to know the experimental result. Till this very moment, on the scale of GREATE UNIVERSE, the cat remains half-alive and half-dead simultaneously. If we recognise only individual consciousness (or, in a narrower sense, only human), then, before the origin of man, the whole Universe should have been in the virtual state of Schrödinger's cat. And man *per se* could not have appeared in this virtual world as reality. And, if it is necessary to be reflected in consciousness to make reality and transfer to the «present», then what does Consciousness differ from Reality? And are the the space and time not the derivatives of Consciousness proper?

The teleological systemic approach to *Universum* and *Evolution* as *Development*, but not as formal tautology of the forms of still (frozen) content could explain both time arrow and reduction of wave function. Teleology can be determined as aimful casualty, as the aim is directed to the future and it selects what is «needed» for this future in the present. These are not objects that form wholiness at random, but it is wholiness that groups elements into new systems for its own «concern». As for such phenomenon as life and reason, herewith it runs not about a transition from order into chaos, but *vice versa* - the occurrence of a higher level order. By the way, at lower levels, there are no grounds for creating the animate from inanimate and the reasonable from automatic. Nevertheless, the desire to make a finite strictly deterministic theory initiating development (inevitably denying the former axiomatics and casualty proper) is still topical. As M. Mesarovich once acrimoniously said: «Teleology is a dame that no biologist can do without, but one is ashamed to be in public with her».

The attempts to confine Evolution to random processes do not sustain any serious criticism. There is an interesting dialogue between a biologist Darwin and von Neumann [4], a famous mathematician. The mathematician showed the biologist from the window of his room and said:

Can you see a wonderful white house over there on the hill? It came into view at random. For millions of years, geological processes have been forming this hill, trees were growing, drying up and decaying, and then growing up again. Then the wind covered the top of the hill with sand; probably some volcanic process brought stones there and they piled up on each other in a certain order. This is the way it was going. Naturally, during the history of the Earth - due to these random unordered processes - something other was coming into view all the time. But, once, in so much time, there appeared this house, then it was inhabited and now people live in it.

This explanation strongly embarrassed the biologist. It is known from astrophysical observations that the time of the Universe (not going about our planet) is maximum several milliards of years. No one has ever managed to squeeze a chain of little probable events - into such a «short» period - that have led to the origin of highly developed living forms on Earth. In all cases, when we deal with similar phenomena - whose origin's probability does not go with any serious theoretical grounds - we look for an *external non-random factor connected with the realisation of some new principle or somebody's desire*. For instance, having found many queerly badly chipped stones, we do not explain it by the thing that they randomly rolled down in it from a top and got so successfully and evenly polished on the edges when rolling. First of all we suppose that prehistorical people lived here and that some need made them polish these stones on some *purpose*. The presence of *non-randomised aimful component,* in this phenomenon, immediately explains such a rare «natural abnormality». To realise development, no energy consumption is needed; it is sufficient to control probability forming aims.

The natural-scientific principle of entropy increase can explain, well, how monkey could have developed out of man, but not on the contrary. Even darwinists implicitly refer the cause of a creative act to some external «meta-subject» saying that Nature and Evolution *created* us. The principle of entropy increase (growth) is also teleological as some others, *e.g.* the principle of the least effect. It is due to this that they adapted to physics so problematically.

The probability theory allows us to calculate events and states probabilities with any precision. However, it does not say anything about how a choice of some certain state proceeds during testing or observation. In classical physics, we could hypothesise that this choice is really a consequence of the effect of classical type factors, which, *at least in principle,* could be considered and a concrete realisation of a *random process* [46] could be determined. In quantum mechanics we, finally, came across a truely random event which realisation, *in principle,* cannot be predicted. By the way, coming across random (occasional) events in their life, people - practically - never ask a question like: «how did it happen that I won or lost in the lottery, *etc.*?» Usually, the question sounds like: why did it happen to me, why did I deserve it or what am I to blame for? It is not the question about the meaning, but the one about the essence of an event, and it is addressed not to object mechanisms but to the Subject (Nature, God) and his «motives». In such a setting, the question really has sense, as it correlates with «aimful vectors» (needs/desires) of the meta-subject, *i.e.* with the principles of the present reference system. In subject representation, we already deal not with the meanings of signs, but with their essences (gists). And a sign's essence cardinally changes in each context and each reference system. The point of why our system's principles are like these already refers to *Universum* as *meta-meta-subject.*

BELL INEQUALITY AND EPR PAIRS

> *Criterion of Einstein-Podolsky-Rosen Reality, Bell inequalities, entangled pairs and their states when measured in different bases, experimental check of Bell inequalities.*

In physics, the approach of studying Reality is conditioned by the specificity of the subject of investigation proper. Studying objective and, in fact, object-object relations determines the initial setting of perception world. The process of describing any phenomenon implicitly includes its spatial-temporal localisation. The stage of dividing the reality into objects proceeds without its realisation. But the possibility of experimental proof of our representations is a wonderful peculiarity of physics. If suppositions based on «common sense» contradict to facts, then the facts remain.

The historical aspect of quantum mechanics becoming is very important to understand its problems. Thus, discussions of Einstein and Bohr [43] about the completeness and concordiality of this theory, on the one hand, led to its recognition as an instrument that adequately describes Reality, and, on the other hand, it turned that the gist of this theory cannot be understood in the traditions of classical physics. Let us consider the contribution of A. Einstein, V. Podolsky and N. Rosen (further, we will use the abbreviation EPR for short) published in 1935 [44]. In it, the authors suggested a mental experiment which doubted the completeness of quantum mechanics as a theory that pretends to be complete in Reality description. The basic point of the paper was the postulating of two quite obvious statements. The first one reads: «...If, without any system's exciton, we can significantly predict (probability equal to one) the quantity of some value, then there exists the element of physical reality corresponding to this physical value [43]». In this postulate, it is implied that Reality is independent from the act of perception, *i.e.* it is objective. Besides, it is asserted that reality can be predicted univocally provided at the completeness of the object description. Further on, this point will be analysed in more detail. The second postulate got its name of «local realism» and, in fact, it is a specification of the position of the present description way as object. Its essence consists in the thing that an event that happened in one place cannot be the cause of an event in the other, until the signal about the first one has been transmitted over there. Thus, the accordance of events in the world proceeds with an extreme velocity of information or a signal's transmission - that of light velocity. These postulates quite definitely formulate the principles of object description of reality. Based on them, an experimental solution was proposed.

---

[46] The word-combination «random process» is in the inverted commas as it is not random essentially in such understanding, as it is based, more likely, on our non-wish to get to the truth, or on the inexpedience of precise awareness of the present event.

We already know that the quantum system is determined as «entangled pairs» where states are correlated. Using «object» terminology, measuring a property of one particle is said to allow us to get the information about a property of the other connected with it. Based on the suppositions made, one can assert the thing that properties correlation will be absent in case when the temporal interval between both particles measurement is less than the time required for the signal transmission of spatial interval between them. In physics, the laws of saving allow us to conduct such an experiment. During the decay of a system, new particles - in total, are to have the same complete set of properties as that of the initial system which may form. So, two gamma-quants, which take away the impluse, energy and spin of «disappeared» particles, form during electron and positron annihilation. If an electron-positron pair was in rest (it is always possible to transfer to the mass centre system), in the so-called singlet state $(\downarrow\uparrow)$ - when the total spin projection onto some axis is equal to 0, the formed photons totally have a zero impulse and spin, and the corresponding total energy. As spin operators onto two orthogonal directions are non-commutating, it is possible to check the supposition by measurement. The similar experimental setting was proposed by D. Bohm which later carried out.

However, at the beginning, it was necessary to formulate the point - the answer of which would univocally recognise either the reasonableness of EPR approach to reality description or quantum mechanics, clearly. In 1964, Bell [45] proved the theoremes known as Bell inequalities in scientific papers. These inequalities may be formulated in different ways [43].

Let an object be characterised by three values A, B, C being equal to values ± 1. In quantum mechanics, non-commutating operators **A**, **B** and **C** connected with dynamic variables α, β, γ (*e.g.* spin projections onto axes x, y, z), may correspond to these values. Suppose the particle simultaneously has all characteristics α, β, γ. Then considering the ensemble of identical particles and having indicated α^+ case - when the variable is equal to +1, and α^- - when it is equal to -1, we will obtain:

$$N\left(\alpha^+\beta^+\right) = N\left(\alpha^+\beta^+\gamma^+\right) + N\left(\alpha^+\beta^+\gamma^-\right)$$

where N - a particle number with corresponding properties. From equalities:

$$N\left(\alpha^+\gamma^+\right) = N\left(\alpha^+\beta^+\gamma^+\right) + N\left(\alpha^+\beta^-\gamma^+\right)$$
$$N\left(\beta^+\gamma^-\right) = N\left(\alpha^+\beta^+\gamma^-\right) + N\left(\alpha^-\beta^+\gamma^-\right)$$

it vividly follows that:

$$N\left(\alpha^+\beta^+\right) \le N\left(\alpha^+\gamma^+\right) + N\left(\beta^+\gamma^-\right) \tag{3.29}$$

That has been one of Bell's inequalities is.

Now let us find out what quantum mechanics predicts when measuring an entangled pair (also called EPR pair), *e.g.* photons in two different bases. We choose the following state for EPR pair: $|\Psi_{12}\rangle = \frac{1}{\sqrt{2}}\left(|00\rangle + |11\rangle\right)$, where 0 and 1 indicate horizontal and vertical polarizations (probability amplitudes are equal for both states). And then we measure photons in the bases turned at an arbitrary angle to each other. For the bra-vector, basis transformations are the following:

$$\langle 0_\alpha| = \cos\alpha\langle 0| + \sin\alpha\langle 1|$$
$$\langle 1_\alpha| = \cos\alpha\langle 1| - \sin\alpha\langle 0|$$

We calculate the probability of amplitude of spin particles state's measurement for two bases in general. Focussing on the variant of obtaining state «0» in each of the bases. It may be formulated as follows:

$$\left\langle \Psi_{\alpha\beta} \mid \Psi_{12} \right\rangle = \left\langle 0_\alpha \left| \left\langle 0_\beta \left| \left(\frac{1}{\sqrt{2}} \left(|00\rangle + |11\rangle \right) \right) \right\rangle \right. = \frac{1}{\sqrt{2}} \cos(\alpha - \beta)$$

Thus, the probability of obtaining such a result is:

$$p\left(0_\alpha, 0_\beta\right) = \frac{\cos^2(\alpha - \beta)}{2} \tag{3.30}$$

Analogously, it is possible to present the probabilities of obtaining the rest of the states combinations:

$$p\left(0_\alpha, 1_\beta\right) = \frac{\sin^2(\alpha - \beta)}{2} \tag{3.31}$$

$$p\left(1_\alpha, 0_\beta\right) = \frac{\sin^2(\alpha - \beta)}{2} \tag{3.32}$$

$$p\left(1_\alpha, 1_\beta\right) = \frac{\cos^2(\alpha - \beta)}{2} \tag{3.33}$$

It is clear that, when angles α and β are equal, the probability of obtaining horizontal («0») or vertical («1») polarization will be equal to 1 on the basis of both particles, the rest turning to zero. In fact, it means on the basis of measurement. Therefore, this condition is always observed for the entangled pair. Now let us norm the probability. For this purpose, we put down the ratio for a complete number of particles:

$$N\left(0_\alpha, 0_\beta\right) + N\left(1_\alpha, 0_\beta\right) + N\left(0_\alpha, 1_\beta\right) + N\left(1_\alpha, 1_\beta\right) = 2N_0$$

$$N\left(0_\alpha, 0_\beta\right)/2N_0 + N\left(1_\alpha, 0_\beta\right)/2N_0 + N\left(0_\alpha, 1_\beta\right)/2N_0 + N\left(1_\alpha, 1_\beta\right)/2N_0 = 1$$

$$P\left(0_\alpha, 0_\beta\right) + P\left(1_\alpha, 0_\beta\right) + P\left(0_\alpha, 1_\beta\right) + P\left(1_\alpha, 1_\beta\right) = 1$$

where $2N_0$ - number of measured entangled particles, the last two lines are total probability of measuring two particles in some state. Herewith, for different bases, the condition of complementarity is realised for measurement probability in particle α basis being in horizontal or vertical polarization (formulas 3.30 and 3.32, or 3.31 and 3.33).

The formula to be checked, according to Bell inequality, - for three arbitrary angles α, β and γ, which non-commutating operators correspond to - is derived from formula (3.30) with a simple replacement of ($+$) indexes, *e.g.* by «1»; and ($-$) by «0»:

$$N\left(\alpha^+\beta^+\right) \le N\left(\alpha^+\gamma^+\right) + N\left(\beta^+\gamma^-\right)$$

the same may be written in probability terms as

$$P\left(1_\alpha, 1_\beta\right) \le P\left(1_\alpha, 1_\gamma\right) + P\left(1_\beta, 0_\gamma\right)$$

Let us set angles α, β, γ to be equal to 0, 30 and 60°, respectively, and, having calculated spin measurement probabilities - according to predictions of quantum mechanics - we substitute them in Bell inequality:

$$\frac{\cos^2(\alpha-\beta)}{2} \leq \frac{\cos^2(\alpha-\gamma)}{2} + \frac{\sin^2(\beta-\gamma)}{2},$$

$$(\alpha-\beta)=(\beta-\gamma)=30° \Rightarrow \frac{3}{8} \leq \frac{1}{8}+\frac{1}{8}$$

Due to the aroused contradiction, it follows that these predictions are incompatible.

The fist experiment was carried out in 1974 in France. Proton spin projections onto three different directions were set as variables α, β, γ. However, the finest experiments with EPR-photons were performed by Aspect and co-authors [37, 38]. In these experiments, one of the analysers changed its polarization angle for the time interval, shorter than it was required for measuring spins of both photons. Hence, herewith, any possibility to accord measurement results to EPR supposition was excluded. In the experiment the effect of the turn was achieved by photon acousto-optical interaction with still ultra-sonic wave in water.

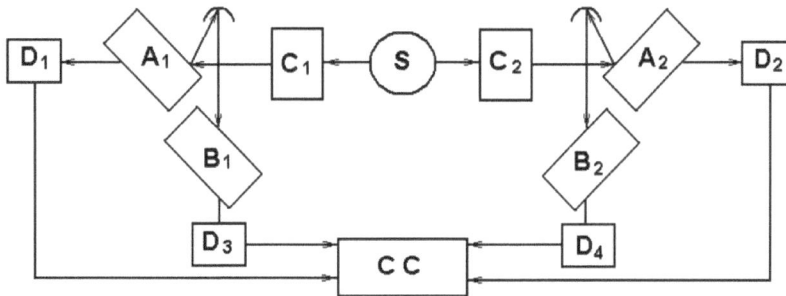

Figure 3.22. The experimental scheme: S - EPR source, D - detectors, A (B) - polarizers, C - switching devices, CC - coincidence circuit.

For this thing, before contacting analysers (Fig. (**3.22**)), photons pass through the turning device (rotator) C $_1$ on the left and C $_2$ - on the right, where they interact with the ultra-sonic wave. Herewith, one photon passes through C $_1$ (second - C $_2$, respectively) without diffraction if still the ultrasonic wave amplitude is zero, and it diffracts at angle 2 θ_B (where θ_B is Bragg angle) if the amplitude is maximal. If light does not diffract, it contacts analyzer A $_1$ (A $_2$), and if it does, it contacts analyser B $_1$ (B $_2$). The experimental geometry is chosen so that time Δt of photons l distance coverage from the source to the turning device, equal to $\Delta t = l/c$, would be more than the «rotation» time. This way, the events of phototons transmission of left and right analysers are separated by the space-like interval. Devices C $_1$ and C $_2$, left and right, do not correlate with each other when operating. The results of the conducted experiment contradict to Bell inequalities and confirm the fairness of quantum mechanics. It turned out that photons spins are completely correlated; no information exchange is required for the «accordance» of particles «pair» behaviour.

At the beginning of the paragraph it was mentioned that EPR suppositions are, in fact, a check of object description's applicability to quantum systems. The negative result is suggestive of the thing that such an approach is impossible. Right is the consideration of a quantum phenomenon in terms of the wholistic state beyond spatial-temporal categories.

- *The simplest variant of Bell inequalities and the corresponding experiments, being statistically significant selections, were analysed above. Presently, entangled three-element systems, for which quantum mechanics has significant predictions contradicting to local realism, have been described. They have been proved in experiments with photons, and those to whom it may concern may get acquainted with them, e.g. [46].*

SEMIOTIC ASPECTS IN REPRESENTATION OF PHYSICAL PROCESSES

Mechanistic process and development, evolution and problem of choice, representation of phenomenon in space-time and Hilbert space, space-time and spectral windows, indeterminacy principle in spectral analysis, physical reality and selection system.

In the concept «process», it is possible to point out two aspects: *realisation of certain content in a different form,* which correlates with *mechanistic motion,* or *a change of this very content,* which correlates with *Development.*

Mechanistic process may be set in two ways: as a sequence of forms of a certain class (quality), whose carrier is object space [47], or by direct indication of *modification rules* of some initial form that set this sequence of forms (*i.e.* the content of a process or calculation procedure). The sequence of forms of a class (quality) determines some process expressing some concrete content. The rule of forms ordering (sequence) of a class (quality) is the content of this process. The rule of forms ordering (sequence) of a class realises the content of this process. For example:

$$1 \to 3 \to 5 \to 7 \to 9 \to ... \to (2n+1) \to ...$$

$$(a+b)^2 \equiv (a+b)(a+b) \equiv (a+b)(b+a) \equiv a^2 + 2ab + b^2 \equiv ...$$

If the sequence of forms of the present class is initiated by a finite operator (particular relevance, rule of output) then the process expresses the *finite content.*

Thus, *the content is expressed in a concrete process* (sequence of forms). There is no other way of content's existence in the space-time. Obviously, the philosophers mean the same when speaking about the indestructibility of motion.

If we turn to the origins of our knowledge, then any science describes feelings, their interconnections and tries to predict them in various conditions [48]. In physics, these predictions are ususally expressed as some function or equation (*e.g.* $x = f(t)$ or $i\hbar\dfrac{\partial \Psi}{\partial t} = \hat{H}\Psi$). Anyway, pointing out the content of a phenomenon from wholiness, we also have to describe it as wholiness in the whole space-time. Herewith, time is a parameter; it does not «flow» anywhere, the past and the future are not outlined. Let us consider the example from electrodynamics.

As is known, Maxwell equations permit solutions not only as delayed potentials $A(r,t) \sim f(t - x/c)$, but also $A(r,t) \sim f(t + x/c)$. Unlike delayed potentials, outrunning ones are accepted to be a non-physical solution because they correspond to the spread of signals reverse in time (from the future), and they are usually dropped. But, in 1945, R. Feynman and J. Wheeler [47] formulated the hypothesis that signals from the future may be Reality, it is just necessary to consider their interference correctly. R. Feynman and J.

---

[47] Space is a logically thinkable form (or structure) that serves as medium in which other forms and these or that constructions are realised. In modern mathematics, space is determined as a multitude of some objects which are named its dots; these may be any geometrical figures, functions, states. Considering their multitude as space, only those properties of their totality, which are accepted on the definition, are taken into account. Relations among dots multitudes determine spatial geometry. The basic properties of these relations are expressed in adequate axioms. Actually, space determines the possibilities (within present properties) of constructing these or that objects and setting concrete processes and states.

[48] Heinsberg - «It is worth noting that the function of probability does not describe the temporal sequence of events proper. It characterises the tendency of an event, its possibility and our knowledge about the event. The function of probability is connected with reality only when one relevant condition is carried out: new observations and measurements are necessary to reveal a certain system's property [4]».

Wheeler needed this hypothesis to resolve the paradox of infinite energy of radiating electron self-effect. Later they refused to accept this idea.

Professor J. Cramer [48] used R. Feynman and J. Wheeler's idea in a quite an unusual way. Not claiming to create a new theory, J. Cramer made up an explanatory scheme for quantum radiation processes - Transactional Interpretation of Quantum mechanics (TIQM). According to his words, this scheme is just the interpretation of quantum mechanics aimed at alleviating the perception of quantum paradoxes including that of EPR described above.

The basic idea of TIQM consists that each elementary radiation act may be presented as «simultaneous» appearance in the space-time of the line on the wave time from the source and a reverse wave from the detector. J. Cramer calls it «out-of-time/timeless 4-dimensional description». He calls the straight wave from the source «wave-sentence» and the back-wave from the detector - «wave-confirmation».

Amplitudes of both straight/direct and back waves are equal to $\frac{1}{2}$ of the usual (direct) radiation wave from the source to the detector observed in the experiment. Herewith, waves phases accord so that there is no electromagnetic radiation field till the radiation moment in the source and after the absorption moment in the detector: waves interfere so that their total is equal to zero in these fields. At the same time, in the intervals between radiation and absorption acts, they interfere so that their sum is equal to the usual wave - the one observed in the experiment.

Absolutely, all these sound logically harmonic, but it does not add anything to the understanding of the thing how that the choice of one of the possible transactions is realised. TIQM evades the point.

Where and when is the choice realised? It is the key question that is in the way of TIQM's becoming a model, but not the interpretation. If the straight (direct) and back-waves really exist at the radiation moment proper, then for *e.g.* in case of EPR-experiment, it means that each of radiated photons has already certain polarization. In other words, the collapse of wave functions occurs not at the moment of photons detection, but right after their separation.

Obviously, the very concept of measurement is illegitimate in quantum formalism, as it already presupposes some objective existence of a value which we still do not know. Moreover, physical values used in quantum mechanics (spin, trajectory, impulse) also imply the objective existence of their values which can be measured, because, otherwise, it is impossible to realise their object representation. And the ritual enunciation appear only at the moment of measurement which does not save things, and it is just a groundless mantra. Remember that objects are constructed as a totality of descriptors with certain intensities on a mental map. However, objects are pointed out from a phenomenon which expresses some wholistic (finite) contents as some constants of functional dependence reflect this content as a process. For instance, the equation of «electron trajectory» in the magnetic field includes such parameters as charge, spin, velocity, *etc.* which may be ascribed to the object «running» over this trajectory. Mental «decomposition» of a phenomenon's whole content and representation of this phenomenon as a process originate such notions as object, space and time. Classical representation of an object as a vector in the space of properties, in quantum mechanics, is replaced by the description of a phenomenon as a whole (*i.e.* all conditions determining process). In the Hilbert space, the function is representable as a vector, and it is not the vector of object presentation, but the one to describe all the phenomena as a whole.

Obviously, to represent *properties* and their *intensities*, we have to identify the type of functions *S(t)* that create feelings, their field of setting and changes. Therefore, an immediate *S(t)* value does not carry any information for an individual [49]. For instance, to discern a sound, as a minimum, it is necessary to point out

---

[49] It is obvious that neither spatial dot nor the temporal (moment) can be perceived as they do not have any differential traits.

a tone, *i.e.* carry out a signal's spectral processing with a continuity of at least two periods [50]. Thus, any feeling is only a seemingly immediate act which implies the process of a signal's «convolution» of some continuity Δt in time. The impression of a feeling's [51] immediateness occurs due to a transfer from an unrealisable *signal* to a deliberate *sign* which is *a qualitative transfer* (immediate act of a sign's appearance in consciousness [52]. The sign implies the whole content that cannot change gradually (you cannot be «a bit pregnant»). A change of a form, when some process is realised, may be continuous, but a change (occurrence) of content is always a «jump».

We describe the perceived process *S(t)* as a function. To make up *a sign (or marker) of a process as a whole*, it needs to be «convoluted» in time, which was discussed above. Spectral transformations may be considered as an example. Every function may be either *strictly harmonic* or non-harmonic. In the last case, it is possible to describe it with an analytical expression, in particular, a Fourier set or integral, *i.e.* to present it as a sum of harmonics [53]. Then, the amplitude will also be a sum of coherent *elementary standard harmonics, i.e.* discrete. Whence it follows that we have no right to use the presentations of an *immediate* amplitude or a frequency which changes in time, as they both automatically originate other harmonics. With such an approach, there is a number of problems having a number of suspicious analogies with problems of quantum physics.

Let us consider the impulse of continuity T (from 0 to T). Its spectrum is presented by the Fourier integral. At some frequency ω, it has the harmonic constituent defined on the *whole infinity of temporal axis* (all the past and future). In fact, it means that the harmonic constituent (component) *already existed in the impulse before the occurrence of the impulse proper, i.e. it is a vivid violation of the law of causality*. This means: if we pass on to the wholistic (non-temporal) Reality perception, *violations of causality occur inevitably*. It is possible to show that, having summed up these constituents for any moment of time $t < 0$, we will get zero, *but not for $t > T$*.

Herewith, it is possible to understand how *experience of time*, which is absent there on the idea, may occur in «limited» Consciousness. Instead of considering the spectrum of the *whole signal* [54] (*i.e.* for infinite «time»), we are limited by its spectral decomposition over *sequences of finite intervals*. It is clear that at different temporal sites, a signal's spectra may be different, *i.e.* there appears an ordered (indexed) *sequence* of different signs. The essence of such temporal variability representation is in the *non-complete Fourier transformation*, as impulse should be considered on the *whole infinity of temporal axis*. Actually, we «cut» *S(t)* into pieces in the temporal axis or (which is equivalent) we pass it through the spectral window in frequency representation, *i.e.* we *filter* it. But if we expose a «cut-out» impulse to the narrow-band filter, then it will pass only some part of an impulse spectrum with the filter-determined changes of amplitudes and phase shifts of constituents. But the higher the *corpulence* of the contour is, the higher its *lag* is; so more time is required for any change of fluctuation amplitude in the contour. It results that when we

---

[50]Nyquist frequency.

[51] Concrete feeling is already differentiated from other feelings in its quality and intensity, and it is a *marker-sign* of some *i*-process and its intensity as integral characteristics of function $s_i(t)$ for time Δt. Temporal quantums of Δt perception are automatically regulated by adaptive mechanisms of setting to relevant signals, *e.g.* settings to informative sonic frequency.

[52] Realising a meaning (or sign perception) is a qualitative change and really, an *immediate act*. Unlike *process*, meaning does not exist in space-time. Absolutely, sign implies information; but physicists do not search these signs, *e.g.* on electron, to understand «where» and «how» the rules of its behaviour in electromagnetic fields or in the interaction with other objects (in essence, it is this way that the *psychophysical problem* is formulated) are recorded. It is obvious that this knowledge belongs to the whole Universe as a totality and cannot be locally presented (it is not present in a spatially localised object). Laws of physics «belong» not to *concrete electrons* and they are not put down in them as behavioural programs in «brains» or any other mechanisms; they just only *realise their content as processes* through these object forms, as the presentation of content in space-time is unthinkable in the way other than processes of information (regularities) translation.

[53] Spectral presentations, judging by all, play a considerable role in our feelings already because coding intensities of all stimuli is realised in neurons as frequency.

[54]Of the whole process that expresses some finite content.

decrease Δt of time quantum, we increase the filter's corpulence and error in the evaluation of amplitude and the signal's energy. It is equivalent to the ratio of indefiniteness in quantum physics: $\Delta E \, \Delta t \sim h$. It follows from here that rigidness of a process is defined not only during perception, but, in a sense, it is originated by it according to quite objective laws. Then it becomes clear why it is not an independent property. By the ratio of indefiniteness, it characterises the parameter of «cutting-off» Δt perception act, as it depends on it *what we are going to perceive*. Thus, continuity of motion in our consciousness, as it was mentioned earlier, is an illusion created by the function of specific brain sections (psychic apparatus).

Feelings are a temporal process. Obtaining information, its interpretation and preservation imply the presence of a semantic space and the object character of knowledge representation to Consciousness. But wholiness, as a minimum, excludes time as *duration*. Only *sequence* is left. The concepts of velocity, duration and distance, in the usual sense, disappear in the subject representation of a phenomenon.

What is «event» in this representation? All «changes» connected with parameter t are identical within a limited content. Function $f(t)$ is defined as a whole and it determines one phenomenon as a whole; it cannot be considered as a process of content evolution. Its spectral presentation $F(\omega)$ mostly corresponds to it. Actually, it is a vector presentation $f(t)$ in the Hilbert space in which we can compare these vectors. However, the measurement of temporal sequence is limited to us by ratio $\Delta\omega \, \Delta t = 1$.

There are considerable differences between temporal and spectral presentations of a process. For instance, even now there is a confusion of two similar but not coinciding notions - *immediate signal frequency* and the *frequency of its spectral constituent* among engineers. In the spectral presentation of a signal as a Fourier set, it is expressed in the sum of harmonics, each of them being a function set on the whole temporal axis and characterised by amplitude, frequency and the initial phase: $A\cos(\omega t + \varphi)$. The signal may be written quasi-harmonically as: $s(t) = A(t)\cos[\omega t + \varphi(t)]$. In the latter, immediate frequency may be determined by complete phase $\Phi(t) = \omega t + \varphi$ through the fluxion $\dfrac{d\Phi(t)}{dt} = \omega(t)$. Although, in the first and second case we deal with frequency; however their properties are different.

Table 3.1: Temporal and spectral presentations.

Immediate frequency	*Frequency of spectrum harmonic constituent*
Is a temporal function.	Does not depend on time.
Has only one value for the present signal at the present moment of time.	For the present signal at any moment of time, there is a finite, countable or countless multitude of spectral constituents with different frequencies.
It may change during a signal's transmission through the linear chain with constant parameters.	Does not change during a signal's transmission through the linear chain with constant parameters; only amplitudes and initial phases may change.
Cannot be an argument of the transfer chain function.	Can be an argument of the transfer chain function.
Measured with various frequency detectors (discriminators).	Measured with analysers of spectra (sets of filters or resonators) or a rearranging resonator.

Herewith, it is possible to find an analogy to the classical and quantum impulse representation. If, in the first case, the impulse was determined by velocity and mass, in the second one - by the de Broglie wavelength.

In general, discussions of radioengineers at the beginning of the last century, in many things, co-echoed with the problems of quantum mechanics interpretation. For instance, in the 1920 - 30's, some outstanding engineers (including A. Fleming, the inventor of diode [49]) opposed the concept of side frequencies that arose during a signal's amplitude modulation:

$$A(1 + m\cos\Omega t)\cos(\omega_0 t + \varphi_0) =$$

$$A\cos(\omega_0 t + \varphi_0) + \frac{m}{2}A\cos[(\omega_0 + \Omega)t + \varphi_0] + \frac{m}{2}A\cos[(\omega_0 - \Omega)t + \varphi_0]$$

Fleming and his supporters supposed that this transformation was one of many possible mathematical representations and it did not say anything about the real existence of side frequencies. For example, even a simple harmonic signal may be decomposed into the sum of several other signals. *E.g.* let us indicate using $M(t)$, the periodical triangular function of period $T = 2\pi / \omega$, determined on interval $-T/2 < t < T/2$ and periodically continued out of this interval with the expression:

$$M(t) = \begin{cases} 1 + 4t/T, t \leq 0 \\ 1 - 4t/T, t > 0 \end{cases}$$

and function $\cos\omega t - M(t)$ - using $N(t)$ indication. Then according to the definition, $\cos\omega t = M(t) + N(t)$. Is it possible, on this basis, to state the thing that the cosine curve really contains triangular function $M(t)$. The falsity of such an approach was proved during a discussion in 1930. I. Mandelstam, a Russian academician, in particular, noted that side frequency (and any spectral constituent) acquires its *physical reality* as soon as the *selective system, being able to discern it*, is used. For example, it is possible to discern the $M(t)$ constituent from a cosine curve using parametrical filter for which the triangular function is its own function [50]. In quantum mechanics, a measurable value arises with the measurement system capable of discerning it.

EXAMPLES OF QUANTUM GATES

One- and two-qubit quantum gates, impossibility of cloning an unknown quantum state.

Earlier it was established that a qubit's state is reflected in the 2D Hilbert space. Any of its reversible change may be represented as an effect of unitary transformation rotating a qubit state's vector. Linear transformation on the complex vector space may be put down as matrix. Let $\mathbf{M}^*$ be a conjugated transponed matrix $\mathbf{M}$. The matrix is called unitary (it describes unitary transformation) if $\mathbf{MM}^* = \mathbf{I}$, where $\mathbf{I}$ is identity matrix. Thus, quantum computer calculation is the function of operators over the qubits register. Bennet and a number of other authors considered reversible variants of standard computation processes [51].

Let us give some simple examples of useful one-qubit quantum states transformations (gates) and the corresponding matrices. The operator of identity $\mathbf{I}$ retains its initial state: $\mathbf{I}|0\rangle \rightarrow |0\rangle$ and $\mathbf{I}|1\rangle \rightarrow |1\rangle$ its matrix $\begin{pmatrix} 1 & 0 \\ 0 & 1 \end{pmatrix}$. Operator (gate) **NOT** or **X**: $\mathbf{X}|0\rangle \rightarrow |1\rangle$, $\mathbf{X}|1\rangle \rightarrow |0\rangle$ matrix $\begin{pmatrix} 0 & 1 \\ 1 & 0 \end{pmatrix}$. Value and phase change operators (gates): $\mathbf{Y}|0\rangle \rightarrow |-1\rangle$, $\mathbf{Y}|1\rangle \rightarrow |0\rangle$ matrix $\begin{pmatrix} 0 & 1 \\ -1 & 0 \end{pmatrix}$ and $\mathbf{Z}|0\rangle \rightarrow |0\rangle$, $\mathbf{Z}|1\rangle \rightarrow |-1\rangle$ matrix $\begin{pmatrix} 1 & 0 \\ 0 & -1 \end{pmatrix}$

The names of these transformations are commonly accepted. One can also make sure that **Y** is a combination of operators $\mathbf{Y} = \mathbf{ZX}$ and that all these gate-transformations are unitary.

Besides, there are two-qubit operators (gates). *E.g.* gate «controlled NOT» or $\mathbf{C}_{NOT}$ is determined for two qubits in the following way. The state of the second qubit called controlled changes for the reverse if the first one (controlling) is equal to 1 and it remains unchanged in the opposite case. Gate $\mathbf{C}_{NOT}$ may not be decomposed into two one-qubit transformations. Below, you can see the operator's truthfulness and its matrix.

$$\begin{pmatrix} a & b & a & b \\ 0 & 0 & 0 & 0 \\ 0 & 1 & 0 & 1 \\ 1 & 0 & 1 & 1 \\ 1 & 1 & 1 & 0 \\ in & & out & \end{pmatrix} \Rightarrow \begin{pmatrix} \mathbf{C}_{NOT}|00\rangle \to |00\rangle \\ \mathbf{C}_{NOT}|01\rangle \to |01\rangle \\ \mathbf{C}_{NOT}|10\rangle \to |11\rangle \\ \mathbf{C}_{NOT}|11\rangle \to |10\rangle \end{pmatrix} \Rightarrow \begin{pmatrix} 1 & 0 & 0 & 0 \\ 0 & 1 & 0 & 0 \\ 0 & 0 & 0 & 1 \\ 0 & 0 & 1 & 0 \end{pmatrix}$$

It is useful to have a graphical representation of quantum state transformations, especially when several quantum states are combined. The controlled NOT is usually presented in the form of the following as shown in Fig. (**3.23**). The controlling qubit is designated with an open circle, the controlled one - with a cross. Controlled qubits may be multiplied. It is also shown in Fig. (**3.23**) where gate **CC** $_{NOT}$ or «double-controlled NOT» is drawn.

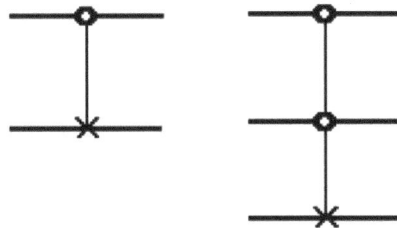

Figure 3.23. C $_{NOT}$ and CC $_{NOT}$ gates.

The double-controlled NOT or Toffoli gate denies the last bit out of three then, and only then, when the first two are equal to 1. The operator matrix is just a record of qubits states at the output after its effect [51]. One-qubit gates are graphically labelled in frames. By the way, two-qubit gates may be presented as a totality of one-qubit. It is possible to present the graphically controlled NOT and double-controlled NOT the way it is shown in Fig. (**3.24**). Using the logical operations and addition on module 2, indicated as « $\oplus$ « ($0 \oplus 0 = 0$, $1 \oplus 0 = 1$, $1 \oplus 1 = 0$), one can formulate the function of operator **C** $_{NOT}$ as a sum of input qubits values. Analogously, gate **CC** $_{NOT}$ is put down using the Boolean cross-operation.

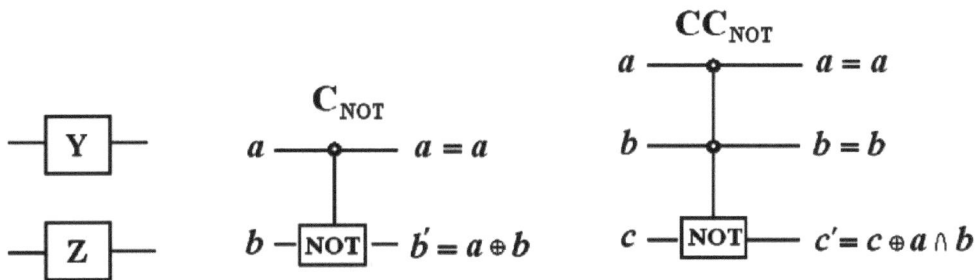

Figure 3.24. One- two- and three-qubit gates.

Hadamard transformation is another important one-qubit transformation defined as follows:

$$\mathbf{H}|0\rangle \to 1/\sqrt{2}\left(|0\rangle + |1\rangle\right)$$

$$\mathbf{H}|1\rangle \to 1/\sqrt{2}\left(|0\rangle - |1\rangle\right)$$

You can easily make up its matrix on your own. Let us give one more record of this operator frequently coming accross in scientific papers:

$$\mathbf{H}|x\rangle = \frac{1}{\sqrt{2}} \sum_{y=0,1} (-1)^{x \cdot y} |y\rangle.$$

Hadamard gate has a big number of important supplements. When $\mathbf{H}$ is used for the qubit in state $|0\rangle$, it makes the superpositional state $\frac{1}{\sqrt{2}}(|0\rangle + |1\rangle)$. It means that measuring a qubit at probability $\frac{1}{2}$ will result in state «0» or «1». If we turn back to the experiment on two slits, then it is adequate to the case when interference destroys if peeped. The initial state of superposition - «transmission of a particle through both slits» - is reduced to the state that corresponds to its detection in one of them. Hadamard gate, applicable to n qubits, generates the superposition of all 2^n states which, binarily, mean a simultaneous record of all numbers from 0 to 2^{n-1}:

$$(H \otimes H \otimes ... \otimes H)|00...0\rangle =$$

$$\frac{1}{\sqrt{2^n}}((|0\rangle + |1\rangle) \otimes (|0\rangle + |1\rangle) \otimes ... \otimes (|0\rangle + |1\rangle)) = \frac{1}{\sqrt{2^n}} \sum_{x=0}^{2^n-1} |x\rangle$$

It denotes the thing that, in the change of n-qubits register, the probability of obtaining any binary set of variables is the same within the interval $0 - 2^{n-1}$. But only one value out of all possible n-local sets will be obtained under the register state reduction. For developing the superpositional state or read-out of all the complete set, many measurements of the similarly prepared register will be necessary, which is analogous to obtaining a clear photofilm picture under the interference of identical particles. The advantage of quantum computer consists in the possibility of creating the superpositional register state carried out by some operator. This property got its name of «quantum parallelism». However, to realise a device's computing power, the use of special algorithms - which would allow us to get a «clear» result - is necessary without a numerous output operation repeat. The way out of the described situation is seen in the very manner of information recorded and read-out. Such peculiarities of the function as periodicity could be reflected by the corresponding operator in the qubit phasal space. For instance, period k may be presented as state $|\psi\rangle = \sqrt{\frac{k-1}{k}}|0\rangle + \sqrt{\frac{1}{k}}|1\rangle$, recorded in each qubit of the register. When measuring n-qubits, the probability of state $|1\rangle$ occurrence will be equal to $1/k$, and it just reflects the period's value.

Having used Hadamard gate for qubits register, it is possible to get state x, corresponding to the binary set from «00...0» to «11...1», simultaneously at the input, and further use of operator $\mathbf{U}_f$ for the input - realising linear transformation over all basis vectors under superposition simultaneously - will allow us to generate results superposition. Thus, it is possible to calculate function $f(x)$ for n-input values of x when using $\mathbf{U}_f$:

$$\mathbf{U}_f \left(\frac{1}{\sqrt{2^n}} \sum_{x=0}^{2^n-1} |x,0\rangle \right) = \frac{1}{\sqrt{2^n}} \sum_{x=0}^{2^n-1} \mathbf{U}_f (|x,0\rangle) = \frac{1}{\sqrt{2^n}} \sum_{x=0}^{2^n-1} |x, f(x)\rangle$$

The power of quantum computer is based on quantum parallelism which allows us, with the help of special algorithms, to calculate the superposition of *all values* for the function of interest from its domain.

Let us mention one more ability of quantum computer. One can show that, using two gates $\mathbf{H}$ and $\mathbf{C}_{NOT}$, can make it possible to make up an entangled pair, the first controlling qubit being superposed:

$$\mathbf{H}|0\rangle \to \frac{1}{\sqrt{2}}(|0\rangle + |1\rangle)$$

$$\mathbf{C}_{NOT}\left(\frac{1}{\sqrt{2}}(|0\rangle + |1\rangle)\right)|0\rangle \to \frac{1}{\sqrt{2}}(|00\rangle - |11\rangle)$$

Generally, entangled states are rather a norm than exclusion for the quantum world. Let us analyse one interesting point on the impossibility of creating quantum state duplicate - the so-called «prohibition for cloning» [51].

Prove it from the contrary, *i.e.* suppose $\mathbf{U}$ is a unitary transformation which can «clone» an unknown quantum state; so for $\mathbf{U}|a0\rangle = |aa\rangle$ all states $|a\rangle$. Let $|a\rangle$ and $|b\rangle$ be two orthogonal quantum states. Then also $\mathbf{U}|a0\rangle = |aa\rangle$ follows from $\mathbf{U}|b0\rangle = |bb\rangle$. Consider superpositional state $|c\rangle = \frac{1}{\sqrt{2}}(|a\rangle + |b\rangle)$. Due to the operator linearity we will get that:

$$\mathbf{U}|c0\rangle = \frac{1}{\sqrt{2}}(\mathbf{U}|a0\rangle + \mathbf{U}|b0\rangle) = \frac{1}{\sqrt{2}}(|aa\rangle + |bb\rangle) \tag{3.34}$$

But if $\mathbf{U}$ is a cloning gate, then $\mathbf{U}|c0\rangle = |cc\rangle = \frac{1}{2}(|aa\rangle + |ab\rangle + |ba\rangle + |bb\rangle)$, which is not equal to (2.34). Hence, our supposition is wrong. Let us underline one point: you cannot clone an unknown state. It is possible to reproduce both states $|0\rangle$ and $|1\rangle$, but they do not have the information more than that a classical qubit has, and you will not make it cloning a superpositional state.

DENSE CODING AND TELEPORTATION

Algorithms of dense coding and teleportation of an unknown quantum state.

One can understand the principle of using simple quantum isolators in two examples - dense coding and teleportation. Experiments that demonstrate these phenomena also allow us to specify some peculiarities of the concept «information». In dense coding - if we use «object terminology» - one qubit from EPR pair is used for coding and transmission of two classical qubits. Such a result is surprising if one thinks EPR pair to be in two particles but not a unified system, as only one information bit may be obtained in measuring one qubit. Teleportation is reverse towards dense coding: two classical bits are used in it to transfer the state of one qubit. Teleportation is surprising because cloning is prohibited in quantum mechanics and, nevertheless, it is possible to transmit an unknown quantum state. Moreover, quantum state teleportation has no spatial-temporal limits because there are no such concepts in the phenomenon.

To realise dense coding operations and quantum state teleportation, quantum systems, called entangled pairs, are used (EPR pairs or, better say, EPR systems). Initial conditions are the same in both cases. Let Alice and Bob have an information and qubit exchange channel. Besides, there is a source of EPR systems. First the system's distribution - or, «classically» speaking - each addressee is sent one of entangled particles - is realised. Note that the state is described by one wave function $\psi_0 = \frac{1}{\sqrt{2}}(|00\rangle + |11\rangle)$. Thus, the quantum system's division is conditional. Assume that Alice received the first particle, Bob - the second one. It will be reflected in the sequence of qubits wave functions record. After the system has been distributed, Alice can realise transformations over her part, Bob - over his own. It is clear that any operation changes the state of the whole system.

Let Alice want to transmit two classical qubits coding numbers from 0 to 3. Depending on the coded number, Alice uses one of transformations **I, X, Y, Z** over her part of the system. The sequence of operations and finite states are presented below. Then Alice sends her qubit to Bob.

$$\psi_0 = \frac{1}{\sqrt{2}}\left(|00\rangle + |11\rangle\right) \rightarrow 0 \rightarrow (\mathbf{I} \otimes \mathbf{I})\psi_0 \rightarrow \psi_0 = \frac{1}{\sqrt{2}}\left(|00\rangle + |11\rangle\right)$$

$$\psi_0 = \frac{1}{\sqrt{2}}\left(|00\rangle + |11\rangle\right) \rightarrow 1 \rightarrow (\mathbf{X} \otimes \mathbf{I})\psi_0 \rightarrow \psi_1 = \frac{1}{\sqrt{2}}\left(|10\rangle + |01\rangle\right)$$

$$\psi_0 = \frac{1}{\sqrt{2}}\left(|00\rangle + |11\rangle\right) \rightarrow 2 \rightarrow (\mathbf{Y} \otimes \mathbf{I})\psi_0 \rightarrow \psi_2 = \frac{1}{\sqrt{2}}\left(-|10\rangle + |01\rangle\right)$$

$$\psi_0 = \frac{1}{\sqrt{2}}\left(|00\rangle + |11\rangle\right) \rightarrow 3 \rightarrow (\mathbf{Z} \otimes \mathbf{I})\psi_0 \rightarrow \psi_3 = \frac{1}{\sqrt{2}}\left(|00\rangle - |11\rangle\right)$$

Initial state   *Transformation*   *New state*

Bob applies operator $\mathbf{C}_{NOT}$ to his «part» - the qubit from the entangled pair using the obtained qubit as the controlling one.

$$\mathbf{C}_{NOT}|\psi_0\rangle = \mathbf{C}_{NOT}\frac{1}{\sqrt{2}}\left(|00\rangle + |11\rangle\right) \rightarrow \frac{1}{\sqrt{2}}\left(|00\rangle + |10\rangle\right) \rightarrow \frac{1}{\sqrt{2}}\left(|0\rangle + |1\rangle\right) \rightarrow |0\rangle$$

$$\mathbf{C}_{NOT}|\psi_1\rangle = \mathbf{C}_{NOT}\frac{1}{\sqrt{2}}\left(|10\rangle + |01\rangle\right) \rightarrow \frac{1}{\sqrt{2}}\left(|11\rangle + |01\rangle\right) \rightarrow \frac{1}{\sqrt{2}}\left(|1\rangle + |0\rangle\right) \rightarrow |1\rangle$$

$$\mathbf{C}_{NOT}|\psi_2\rangle = \mathbf{C}_{NOT}\frac{1}{\sqrt{2}}\left(-|10\rangle + |01\rangle\right) \rightarrow \frac{1}{\sqrt{2}}\left(-|11\rangle + |01\rangle\right) \rightarrow \frac{1}{\sqrt{2}}\left(-|1\rangle + |0\rangle\right) \rightarrow |1\rangle$$

$$\mathbf{C}_{NOT}|\psi_3\rangle = \mathbf{C}_{NOT}\frac{1}{\sqrt{2}}\left(|00\rangle - |11\rangle\right) \rightarrow \frac{1}{\sqrt{2}}\left(|00\rangle - |10\rangle\right) \rightarrow \frac{1}{\sqrt{2}}\left(|0\rangle - |1\rangle\right) \rightarrow |0\rangle$$

After this, Bob measures the second qubit and, as is seen from the table, he gets the state of either «0» or «1» at probability 1. If the measurement result is equal to state «0», then the coded value was 0 or 3; if one, then either 1 or 2 was coded. Bob uses Hadamard operator **H** for the first bit left:

$$\mathbf{H}|\psi_0'\rangle = \mathbf{H}\frac{1}{\sqrt{2}}\left(|0\rangle + |1\rangle\right) \rightarrow \frac{1}{\sqrt{2}}\left(\frac{1}{\sqrt{2}}\left(|0\rangle + |1\rangle\right) + \frac{1}{\sqrt{2}}\left(|0\rangle - |1\rangle\right)\right) = |0\rangle$$

$$\mathbf{H}|\psi_1'\rangle = \mathbf{H}\frac{1}{\sqrt{2}}\left(|1\rangle + |0\rangle\right) \rightarrow \frac{1}{\sqrt{2}}\left(\frac{1}{\sqrt{2}}\left(|0\rangle - |1\rangle\right) + \frac{1}{\sqrt{2}}\left(|0\rangle + |1\rangle\right)\right) = |0\rangle$$

$$\mathbf{H}|\psi_2'\rangle = \mathbf{H}\frac{1}{\sqrt{2}}\left(-|1\rangle + |0\rangle\right) \rightarrow \frac{1}{\sqrt{2}}\left(-\frac{1}{\sqrt{2}}\left(|0\rangle - |1\rangle\right) + \frac{1}{\sqrt{2}}\left(|0\rangle + |1\rangle\right)\right) = |1\rangle$$

$$\mathbf{H}|\psi_3'\rangle = \mathbf{H}\frac{1}{\sqrt{2}}\left(|0\rangle - |1\rangle\right) \rightarrow \frac{1}{\sqrt{2}}\left(\frac{1}{\sqrt{2}}\left(|0\rangle + |1\rangle\right) - \frac{1}{\sqrt{2}}\left(|0\rangle - |1\rangle\right)\right) = |1\rangle$$

First qubit state   **H** *Effect*   *Result*

Finally, Bob measures the resulting qubit which allows him to differentiate 0 and 3, 1 and 2 - as he extracts binary set {00} (it corresponds to the transmission of number 0, {10} to 1, {11} to 2, {01} to 3) - out of two measurements.

Now let us consider the transmission of a particle's quantum state using classical bits and reconstructing the state with the addressee. As the quantum state cannot be coded, the state of the transferred qubit destroys when transmitted. Let Alice send an unknown state of qubit $\phi = a|0\rangle + b|1\rangle$ to Bob over quantum channels.

Just as in the previous state, Alice and Bob - each has one qubit from the entangled EPR system $\psi_0 = \frac{1}{\sqrt{2}}\left(|00\rangle + |11\rangle\right)$. The initial quantum state of three particles may be put down in the common basis as,

$$\phi \otimes \psi_0 = \frac{1}{\sqrt{2}}\left(a|0\rangle \otimes \left(|00\rangle + |11\rangle\right) + b|1\rangle \otimes \left(|00\rangle + |11\rangle\right)\right) =$$
$$\frac{1}{\sqrt{2}}\left(a|000\rangle + a|011\rangle + b|100\rangle + b|111\rangle\right)$$

in which Alice controls the first two qubits, Bob does the last one. Alice applies the «controlled NOT» (**C**$_{NOT}$) and Hadamard **H** operator:

$$\left(\mathbf{H} \otimes \mathbf{I} \otimes \mathbf{I}\right)\left(\mathbf{C}_{NOT} \otimes \mathbf{I}\right)\left(|\phi\rangle \otimes |\psi_0\rangle\right) =$$
$$\left(\mathbf{H} \otimes \mathbf{I} \otimes \mathbf{I}\right)\left(\mathbf{C}_{NOT} \otimes \mathbf{I}\right)\frac{1}{\sqrt{2}}\left(a|000\rangle + a|011\rangle + b|110\rangle + b|101\rangle\right) =$$
$$= \frac{1}{2}\left(|00\rangle\left(a|0\rangle + b|1\rangle\right) + |01\rangle\left(a|1\rangle + b|0\rangle\right) + |10\rangle\left(a|0\rangle - b|1\rangle\right) + |11\rangle\left(a|1\rangle - b|0\rangle\right)\right)$$

After that she measures the first two qubits and obtains one of the states corresponding to $|00\rangle$, $|01\rangle$, $|10\rangle$ or $|11\rangle$ with equal probability $\frac{1}{4}$. Depending on the measurement result, the qubit quantum state of Bob's EPR system is randomly projected into one of states, respectively $a|0\rangle + b|1\rangle$, $a|1\rangle + b|0\rangle$, $a|0\rangle - b|1\rangle$, $a|1\rangle - b|0\rangle$. Then Alice sends her measurement results to Bob as two classical bits. Note that, when measured, the initial state of the original qubit ϕ she sends to Bob destroys. This loss of the original initial state is the reason for the thing that teleportation is not in cloning prohibition.

When Bob receives two classical bits from Alice, he comes to know how to modify the state of his left-over «half» of EPR-system to get the initial quantum state corresponding to the one of Alice:

Received bits		State		Coding
00	→	$a|0\rangle + b|1\rangle$	→	**I**
01	→	$a|1\rangle + b|0\rangle$	→	**X**
10	→	$a|0\rangle - b|1\rangle$	→	**Z**
11	→	$a|1\rangle - b|0\rangle$	→	**Y**

Now Bob can reconstruct the initial state of Alice's qubit ϕ, using the corresponding decoding transformation for his qubit (operator in the last column of the table).

Let us stress out two important circumstances: manipulation with «one part» (division being conditional) of EPR system changes all the state right as a whole, the spatial-temporal description «inside» the system being unreasonable. When analysing the EPR pair from «outside», it is possible to «extend» it over any distance related to some external co-rdinate system. However, using the operator immediately changes the whole state, and the process of measurement in one reference system leads to an uncontrolled random occurrence of some result. To make this result information for the other reference systems, it is necessary to transmit a signal according to the code accepted by both the parties. Under teleportation, there is no aberration of SRT which just implicitly asserts the thing that information is the result of an observer's activities. It turns out that the world's accordance *per se* has no limits, and the world's description as an information process is limited within certain frames.

It is only left to add the thing that teleportation and dense coding were first experimentally realised in 1997 [52].

Now let us tackle the physical realisation of some operators. Let us realise Mach interferometer having two ordinary and two semi-transparent mirrors. The incoming photon spreads after transmitting the first semi-transparent mirror to ordinary mirrors from which it is reflected, and then it contacts the second semi-transparent mirror (Fig. (**3.25**)).

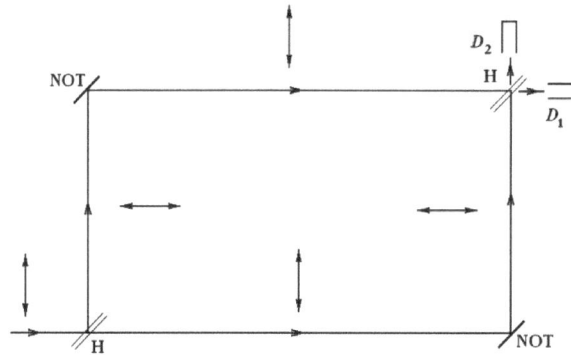

Figure 3.25. Mach interferometer.

We will use linear-polarized radiation, *e.g.* laser. If we cover the lower horizontal arm with a shutter, then detector D_1 will function and the photon with a spin corresponding to that of the input one will be registered. If we close the upper horizontal arm, then detector D_2 will function and the photon with an orthogonal spin will be registered at the output. If all the ways are open, only D_1 will function. The thing is that the usual mirror of a photon is the realisation of «NOT» operator, semi-transparent being for Hadamard operator. Let the photon input spin be $|0\rangle$, then, after the following transformations of **H**, **NOT**, **H**, it is not difficult to get what will also have state $|00\rangle$ at the output.

QUANTUM CRYPTOGRAPHY

Key Formation Algorithm

In the presentation of the material on quantum computations, one cannot avoid the point of quantum cryptography. It is, first of all, connected with a wide interest - and, what is most important, financial support of quantum computer research - which began in 1994 since the publication of Peter Shor [53] on quantum algorithm for factorization (decomposition of a number into prime factors) into polynominal time related to numeric value. To understand the relevance of this result, it is necessary to know that, using modern computer nets functioning on the latest algorithm, it will take about 10 milliard years to decompose a 400-digit number. This task may be resolved for only several minutes using the quantum computer and Shor algorithm. It is the temporal factor that serves reliable guarantee of present codes - based on coding a message by prime factors - security (see Part II «Classical computer»).

What new can quantum cryptography propose? It is a new key transfer way which provides its full secrecy excluding the access of an information code to an outsider when transmitted. The algorithm of quantum key distribution, using polarization photons, was suggested by C.H. Bennett and G. Brassard [54]. By the present, this transmission protocol has already been realised at the distance of some tens of kilometres and it is practically being brought to life as a commercial project. Let us analyse the principle scheme of this transmission.

Let there be an impulse laser capable of emitting single photons and a device (Pockels cell) capable of changing their polarization related to two bases (Fig. (**3.26**)). Settled is to name one basis «direct» (straight) and to indicate it as $\oplus$, and the one turned to it at $45°$ - «skew» designated as $\otimes$. We will think the horizontal state of photon spin $\langle\leftrightarrow\rangle$ in direct basis to be «0», the state $\langle \updownarrow \rangle$ - «1» and that of photon spin $\langle \nearrow \rangle$ in the skew basis - «0», and state $\langle \searrow \rangle$ - «1». It is clear that Alice and Bob can exchange information transmitted in the binary code in this or that basis. For example, in the direct basis, Alice sends a series of impulses $\langle \leftrightarrow \rangle, \langle \updownarrow \rangle, \langle \leftrightarrow \rangle, \langle \leftrightarrow \rangle, \langle \updownarrow \rangle$ which correspond to binary number «10110». To read out the message, Bob uses a polarization light-divider that transmits vertical polarization and reflects the horizontal one when set to reception in the direct basis. Signal read-out is realised by one-photon detectors, as it is shown in Fig. (**3.26**). The countdown in detector D_0 (D_1) means that Alice sent the signal corresponding to value «0» («1»). In this case, Bob is said to carry out detection at $\oplus$ -basis.

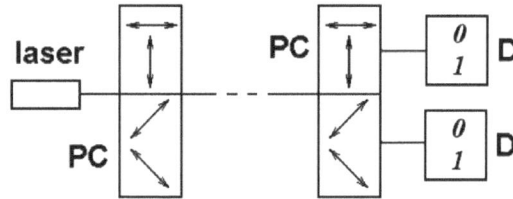

Figure 3.26. A principle reception-transmission scheme.

In Fig. (**3.26**) there is a principle reception-transmission scheme realised by one-photon impulses: polarization is controlled by Pockels cells which allow us to make a choice among 4 possible bases of photon polarization: $\langle \leftrightarrow \rangle, \langle \updownarrow \rangle, \langle \nearrow \rangle, \langle \searrow \rangle$. The polarization light-divider (splitter) (not shown in the scheme) «splits» a signal into two orthogonal components which are detected by the corresponding detectors (receivers).

If we send photons only in one basis, then there always is a probability of information intercept by a third person (in scientific papers it is accepted to believe that filthy Eva does it). In principle, Eva can intercept the impulse, read it out and send the same one on. Using two bases, the probability of correct reception interpretation is equal to $\frac{1}{2}$ if it is unknown on bases each photon should be recepted in. However, it goes about both Eva and Bob. Such a scheme may be used in the solution of key transmission problem where confidentiality is the basic requirement.

Table 3.2: Transmission of the data.

Basis	⊗	⊕	⊗	⊕	⊗	⊕	⊗	⊕	⊗	⊗	⊕	⊗	⊗	⊕
Spin	↗	↕	↘	↔	↘	↕	↘	↔	↘	↘	↕	↗	↘	↔
Value	0	1	1	0	1	1	1	0	1	1	1	0	1	0

Let Alice, randomly choose the basis and value, sends a series of impulses to Bob, such as it is in Table **3.2**. The upper line of this table corresponds to the chosen basis, the second - to the photon spin, the third - to the value of the transmitted number according to the things which were agreed earlier. From his side, Bob receives signals also randomly setting the basis and receives, *e.g.* the sequence presented in Table **3.3**.

Table 3.3: Registration of the data.

Basis	⊗	⊗	⊕	⊕	⊗	⊕	⊕	⊗	⊗	⊕	⊗	⊗	⊕	⊕
Spin	↗	↗	↕	↔	↘	↕	↔	↗	↘	↔	↘	↗	↘	↔
Value	0	0	1	0	1	1	0	0	1	0	1	0	1	0

It is seen from Table **3.3** that where Bob guessed the state of the basis, the values received by him are correct; the rest of the cases values may coincide with those sent and may occur not due to the quantum state superposition before measurement. The same will proceed in Eva's if she tries to intercept the information. After transmission, Alice enumerates the sequence of bases for each transmitted basis over the open connection channel, and Bob confirms either coincidence or divergence. Herewith, both cross out the values that have not coincided on the basis and preserve those coincided. In Table **3.4**, the upper line corresponds to the operation of bases comparison. The upper value is the basis of Alice, the lower one is that of Bob; the next one denotes the result of comparison, the code values not being mentioned; so, the interception of this information by a third person does not give one a numeric key value.

Table 3.4: Key formation algorithm

Check for basis	⊗	⊕	⊗	⊕	⊗	⊕	⊗	⊕	⊗	⊗	⊕	⊗	⊗	⊕
	⊗	⊗	⊕	⊕	⊗	⊕	⊕	⊗	⊗	⊕	⊗	⊗	⊕	⊕
Value	**0**	-	-	**0**	**1**	**1**	-	-	**1**	-	-	**0**	-	**0**
Check for value				**0**		**1**								**0**
				0		**1**								**0**
Key	**0**				**1**				**1**			**0**		

Now, if need be, it is possible to check the information «leakage», *i.e.* to find out if Eva listens over the connection channel or not. For this purpose, Bob transmits the values to Alice from the preliminarily chosen key again, as is shown in Table **3.4**. Let these be the values in columns 4, 6 and 14. If they coincided with Alice and Bob's, as it is shown in the third column of the table, then the connection channel is «pure»; if they did not, then we have to check it. In the final key formation, only those values are kept which were not sounded over the open channel - column 4 of Table **3.4**. It is not worth focusing that, in our example, the key has made so short, technically, as there are no problems to increase its length. Now this key may be used to code messages and send them over open connection channels, as its uniqueness and secrecy are provided by the way of its creation.

DEUTSCH QUANTUM ALGORITHM

Quantum Computer Block-Scheme, Deutsch's Algorithms

Till present, a possible construction of quantum computer has not been discussed anywhere. To get an idea about its operation in a simpler way, let us show the block-scheme [29] in Fig. (**3.27**). The register having qubits, the state of which changes under the impact of controlling impulse-operators set by a program of common computer, will be the basic node of this device.

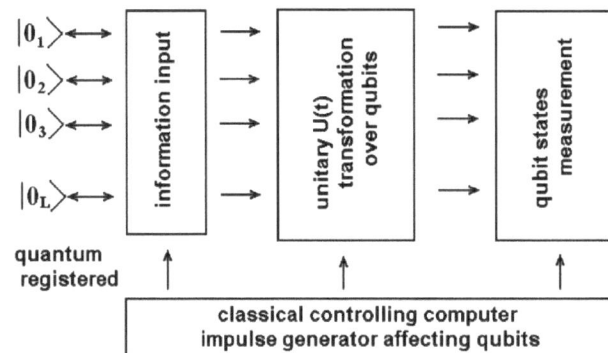

Figure 3.27. A quantum computer block-scheme.

At present, a big number of projects in qubits physical realisation on ions in the electromagnetic trap, superconducting elements - using the donor centre spin nucleus in silicon, spin electron of donor centre, *etc.* - has been proposed. Concrete proposals were analysed in detail in special sources [55-59], and, from now on, we will think that such a device may be fabricated in principle.

Let us consider the simplest quantum algorithm suggested by D. Deutsch [60]. The task is in determining the class of an unknown function which may be either constant or weighed. It means that there are Boolean functions $f(x)$ (for simplicity, we will analyse the case of those single), such as the one that:

$$f_1(0) = 0 \quad or \quad f_2(0) = 1 \quad \text{belonging } to \ the \ class \ of \ constant$$
$$f_1(1) = 0 \quad\quad\quad f_2(1) = 1$$

and

$$f_3(0) = 0 \quad or \quad f_4(0) = 1 \quad \text{belonging } to \ the \ class \ of \ weighed$$
$$f_3(1) = 1 \quad\quad\quad f_4(1) = 0$$

These functions may be set by the following operators:

$$\mathbf{U}_{f_1} = \mathbf{I} \otimes \mathbf{I} \quad\quad \mathbf{U}_{f_3} = \mathbf{C}_{NOT}$$
$$\mathbf{U}_{f_2} = \mathbf{I} \otimes \mathbf{NOT} \quad \mathbf{U}_{f_4} = \mathbf{C}_{NOT}(\mathbf{I} \otimes \mathbf{NOT})$$

Let there be the input operator of two qubits and $|x\rangle$ and $|y\rangle$, the quantum scheme corresponding to the impact of an unknown operator of function $f(x)$ such that $\mathbf{U}_f(|x\rangle \otimes |y\rangle) \Rightarrow |x\rangle \otimes |y \oplus f(x)\rangle$. Examples of such operators were analysed earlier, and a block-scheme of the algorithm as operations carried out consecutively (time axis is directed from left to right) is demonstrated in Fig. (**3.28**). In this case, two qubits are necessary to increase spatial dimensionality to have a possibility to reflect the properties of the function under study with a qubits phase.

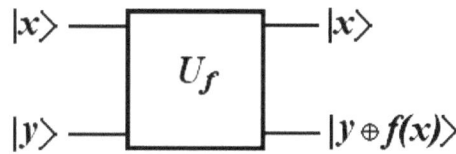

Figure 3.28. A two-qubit quantum gate array performing transformation $\mathbf{U}_f(|x\rangle \otimes |y\rangle) \Rightarrow |x\rangle \otimes |y \oplus f(x)\rangle$

A cycle of contributions on determining the unknown function $f(x)$, after register initialisation (record of state «0») begins with the preparation of qubits $|x\rangle$ and $|y\rangle$ (Fig. (**3.29**)) superposition $\mathbf{H}|0\rangle_x \Rightarrow \frac{1}{\sqrt{2}}(|0\rangle + |1\rangle)$ and $|0\rangle_y \Rightarrow \mathbf{H}(\mathbf{NOT}|0\rangle) \Rightarrow \frac{1}{\sqrt{2}}(|0\rangle - |1\rangle)$. In the function's further calculations for values $|x\rangle$ and $|y\rangle$, the difference of the obtained result - in case of constant and weighed function due to phase difference of these qubits - may be presented as a difference in the amplitude. One has to remember that, when reading out the register state, the information about the phase is lost.

The effect of operator $\mathbf{U}_f$ originates the following register state:

$$\mathbf{U}_{f_i}(|x\rangle \otimes |y\rangle) \Rightarrow |x\rangle \otimes |y \oplus f(x)\rangle \Rightarrow \frac{1}{\sqrt{2}}|x\rangle \otimes (|0 \oplus f(x)\rangle - |1 \oplus f(x)\rangle) \quad\quad (3.35)$$

The expression in round brackets may be extended for two values of Boolean function - «0» and «1» in the following way:

$$(|0 \oplus f(x)\rangle - |1 \oplus f(x)\rangle) \Rightarrow (|0\rangle - |1\rangle) \quad at \quad f(x) = 0$$

$$\left(\left| 0 \oplus f(x) \right\rangle - \left| 1 \oplus f(x) \right\rangle \right) \Rightarrow \left(\left| 1 \right\rangle - \left| 0 \right\rangle \right) \quad at \quad f(x) = 1$$

In a general case, this expression may be presented as follows:

$$\left(\left| 0 \oplus f(x) \right\rangle - \left| 1 \oplus f(x) \right\rangle \right) = (-1)^{f(x)} \left(\left| 0 \right\rangle - \left| 1 \right\rangle \right) \tag{3.36}$$

Substitution of (3.35) to (3.36) leads to the expression like

$$\frac{1}{2} \left(\left| 0 \right\rangle + \left| 1 \right\rangle \right) \otimes (-1)^{f(x)} \left(\left| 0 \right\rangle - \left| 1 \right\rangle \right);$$

having multiplied and regrouped the members, we will get:

$$\frac{1}{2} \left[\left| 0 \right\rangle \otimes (-1)^{f(0)} \left(\left| 0 \right\rangle - \left| 1 \right\rangle \right) + \left| 1 \right\rangle \otimes (-1)^{f(1)} \left(\left| 0 \right\rangle - \left| 1 \right\rangle \right) \right].$$

If the equality $(-1)^{f(1)} = (-1)^{2f(0) \oplus f(1)}$ is to be used, then the above-presented expression may be presented the in following way:

$$\frac{1}{2} (-1)^{f(0)} \left(\left| 0 \right\rangle + (-1)^{f(0) \oplus f(1)} \left| 1 \right\rangle \right) \otimes \left(\left| 0 \right\rangle - \left| 1 \right\rangle \right) \tag{3.37}$$

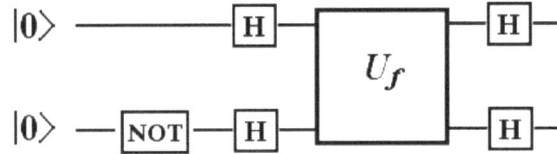

Figure 3.29. Quantum circuit for the Deutsch algorithm.

Now it makes obvious that the first qubit value, after the effect of Hadamard operator, is determined by the phase which depends on the sum of values of an unknown Boolean function at values «0» and «1» - constant and weighed, respectively:

$$f_{1,2}(0) \oplus f_{1,2}(1) = 0 \Rightarrow \pm \left| 0 \right\rangle$$
$$f_{3,4}(0) \oplus f_{3,4}(1) = 1 \Rightarrow \pm \left| 1 \right\rangle$$

It is clear that if the obtained measured output qubit value corresponds to state «0» (phase being lost when measured), then the unknown function belongs to the class of constants, and if it is «1» - to the class of weighed. Thus, we can determine the class in one taking to the unknown function!

QUANTUM ALGORITHMS

Algorithm of amplitude errors correction, Grover and Shor algorithm.

Now let us analyse the errors correction algorithm which is of big importance for full quantum computer functioning. Till 1996, the possibility of creating the quantum computer was problematic in connection with the thing that any real technical device cannot function without breakdowns, and it was unclear how to carry out errors correction. In modern classical computers, the breakdown probability is very small and, moreover, there are effective control algorithms in case of breakdown during a program realisation. There is no special difficulty here as such that correction algorithms are based on the record and comparison of information at any stage of program realisation. In breakdown control, the read-out of current register state is carried out. However, this operation cannot be used in the quantum computer till the end of program realisation, as it completely destroys qubits coherent state. Hence, without a way of errors correction, a real device cannot function normally. Ways of quantum register errors correction were developed in 1996 with the contributions of P. Shor and A. Steane [61, 62]. As an example, let us analyse the principle of the functioning of algorithm that corrects the breakdown of qubit value in case of amplitude error; algorithms of phase error correction are analogous, you can see them in any literary references on quantum computations [29, 46, 51, 63].

Let there be a working qubit $|\psi\rangle$ the state of which we have to control. To realise this operation, we need two additional qubits. These qubits, initially, are in state $|0\rangle$. The graphical scheme of algorithmic functioning is given in Fig. (**3.30**). It is a sequence of $\mathbf{C}_{NOT}$ operations performed over additional qubits, the functional qubit being used as a controlling one. If it is in the state of superposition $|\psi\rangle = a|0\rangle + b|1\rangle$, then, after $\mathbf{C}_{NOT}$ double realisation, we will have the entangled state $a|000\rangle + b|111\rangle$ (left of the scheme, Fig. (**3.30**)). Additional qubits now have the information about the amplitude, adequate to the functional qubit, which is saved in the realisation of control.

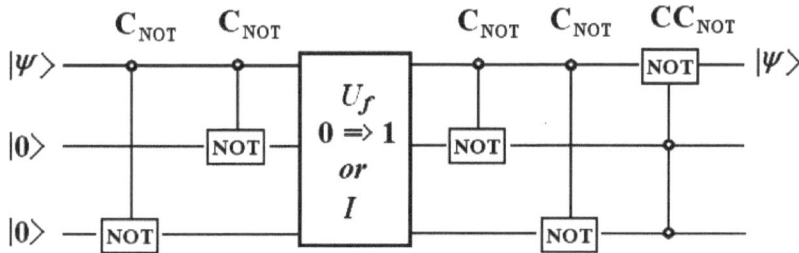

Figure 3.30. Quantum error correction algorithm.

If an error occurred at some moment in the functional qubit (inversion $0 \Rightarrow 1$), then the general state of the set will be: $a|100\rangle + b|011\rangle$ (if it did not, it will remain former $a|000\rangle + b|111\rangle$). The occurrence of error is shown in the centre of Fig. 3.30 as an effect of operator $\mathbf{U}_f$. If, at some moment, we need to control the qubit for errors, then $\mathbf{C}_{NOT}$ operation is carried out over additional qubits using the functional qubit as controlling. As a result, we will get state $a|111\rangle + b|011\rangle$ (state $a|000\rangle + b|100\rangle$ if there was no breakdown). $\mathbf{CC}_{NOT}$ is the last operation of algorithm. But now additional qubits are used as controlling - as is shown in the scheme. As a result, we obtain state $a|011\rangle + b|111\rangle$ ($a|000\rangle + b|100\rangle$), which corresponds to the state of functional qubit $|\psi\rangle = a|0\rangle + b|1\rangle$ for both variants of events development. If the breakdown probability, during computer operation, is so big that a loss of state is possible for the time of program realisation also in additional qubits, it is still early to make calculations with such a device.

Grover Algorithm

Let us consider the functioning principle of Grover algorithm that carries out a search of some element in the unordered database [64-66]. The operation of comparison of each element in the initial database is supposed to be for the value of a function $p(x)$. This algorithm is of big applied importance. However, the quantum analogy does not cause any exponential functional acceleration compared to the classical one. To find a searched value in the classical case, it is necessary to take to the selection of $N/2$ times on the average. It requires high expenses of resources for big arrays (memory, computing means or time). L.

Grover showed that, in the quantum case, it is possible to do with $\sim \sqrt{N}$ addressing on the average, which still accelerates the search process.

Let there be a selection N long having element x_j - such one that the following condition is carried out for probe function $p(x)$:

$$p(x) = \begin{bmatrix} 1, & x = x_j \\ 0, & x \neq x_j \end{bmatrix}$$

There is an n-long register - such one that $2^n \geq N$. The superpositional state of the register having all possible values of argument $|x\rangle$ is prepared at the first stage, and calculations of function $p(x)$ are carried out from all arguments:

$$\mathbf{U}_p |x,0\rangle \Rightarrow |x, p(x)\rangle$$

$$\mathbf{U}_p \frac{1}{\sqrt{2^n}} \sum_{i=0}^{N-1} |x_i, 0\rangle \Rightarrow \frac{1}{\sqrt{2^n}} \sum_{i=0}^{N-1} |x_i, p(x_i)\rangle$$

As a result, in register $|y\rangle$, we will get the state in which the information about the searched element is distributed over all qubits so that probability $p(x_j)$ is different from the rest at value $\dfrac{1}{2^n}$ for the searched element. Earlier it was analysed how the function peculiarity may be reflected by the measurement probability of the value when reading out the output register state.

Changing the phase is made at the next stage. Let there be operator $\mathbf{U}_\Sigma$ which performs the summing on module 2 - such one that: $\mathbf{U}_\Sigma |x,z\rangle \Rightarrow |x, x \oplus z\rangle$. We choose $|z\rangle = \dfrac{1}{\sqrt{2}}(|0\rangle - |1\rangle)$, and x becomes equal to 0 and 1. Then:

$$\mathbf{U}_\Sigma |x,z\rangle \Rightarrow |x, x \oplus z\rangle \Rightarrow \begin{cases} x \oplus z(x=0) \to |z\rangle = \dfrac{1}{\sqrt{2}}(|0\rangle - |1\rangle) \\ x \oplus z(x=1) \to -|z\rangle = \dfrac{1}{\sqrt{2}}(|1\rangle - |0\rangle) \end{cases}$$

changes its phase.

If we use the analogous operator for the searched function $p(x_j)$, it will lead to the phase inversion for the case $p(x = x_j) = 1$, not changing it $p(x \neq x_j)()=0$. Graphically, output register probabilities may be presented for the searched value with such a scheme:

At the next stage, averaging operation is performed to increase the searched element's amplitude by means of an inconsiderable decrease of the rest values amplitude:

$$\mathbf{U}_s \sum_{i=0}^{N-1} a_i |x_i\rangle \Rightarrow \sum_{i=0}^{N-1} (2\bar{a} - a_i) |x_i\rangle \Rightarrow \begin{cases} 2\bar{a} - a_i \approx \bar{a}(x \neq x_j) \\ 2\bar{a} - (-a_j) \approx 3\bar{a}(x = x_j) \end{cases}$$

Graphically, it may be presented as follows:

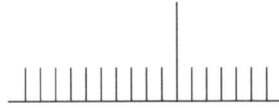

With the repetition of these operations, it is possible to on-grow the searched element's probability amplitude, and it will allow us to detect it, when measuring the register, with quite a high probability (not more than $\frac{1}{2}$, see [29, 46, 51, 63]).

Shor Algorithm

Earlier, the value of Shor algorithms was mentioned regarding the possibility of present codes decoding. To make decoding in the classical computer, actually, sorting out variants on factorisation (search of co-factors), a big N number is used. The main idea of quantum algorithm is based on the thing that the presence of a co-factor leads to the occurrence of the mode that corresponds to it under reverse Fourier transformation. Herewith, discrete transformation in the variant of rapid transformation on base 2 is used as the most matched for computing. The quantum Fourier transformation operator proper, as for the quantum computer register, for number $N = 2^n$ may be set the following way:

$$\mathbf{U}_{QFT}\left|x\right\rangle \rightarrow \frac{1}{\sqrt{2^n}}\sum_{k=0}^{2^n-1}e^{i2\pi kx/2^n}\left|k\right\rangle.$$

A peculiarity of quantum systems is such a phenomenon as interference which reflects the realisation multitude of different properties of a system and it manifests itself when measured. This principle difference between classical and quantum physics allows us to considerably widen a circle of effectively solved tasks. The basic complexity in realisation of quantum computer advantages consists in the creation of an algorithm based on the adequate operators that lead the task to its «interferential» form. It is a serious problem and, probably, it is for this reason that the number of such algorithms is currently not big.

Consecutive description of Shor algorithm requires the formulation of Fourier transformation apparatus, discrete and rapid Fourier transformation. In this connection, we decided to confine it only to a brief note regarding the proper idea proper which is the base of the algorithm. Those to whom it may concern, as for mathematical niceties of its realisation, can take to original [53, 67] papers and manuals [46, 51, 63] in which the algorithm is analysed in quite a detail.

CONCLUSIONS

To wind up the material formulated in parts 1 and 2 of the present contribution, it is necessary to emphasize the following points. Natural to us is the object way of Reality description. Only the understanding of this fact leads to the realisation of the «source» of our feelings, and it allows us to remove a number of problems.

First, it is the so-called psycho-physical problem which initiates the outlook on some transendental «subject world» (like in Platon and Kant) being on-the-other-side of feelings. As subject has no chance to get out of «feelings cocoon», the question about how we receive some «signals» from this «on-the-other-side world» (or how physiological disturbance transfers into a psychological feeling and becomes information) remains in the air - just as the point of the possibility of «objective Reality» cognition and approaching to «true knowledge» presented in this transcendental world. The problem is also in the thing that, within subject-object categories, it is impossible to analyse their «interaction» as an object representation excludes subject (or confines it to object), and the subject excludes object. It breaks Reality into the object world and subject making an unbridgeable gulf between them. And if the object Being is supposed to be primary and the subject - secondary (derivative of this «physical» Being), then it loses its semiotic definiteness and, actually, becomes Absolute that has no right to participate in any process of semiosis as it inevitably rises logical contradictions [55].

Second, spatial-temporal presentation of Reality (accepted as initial), alongside with the limitness of a process's extreme velocity, leads to the principle of locality which was experimentally disproved by quantum mechanics (EPR effects, quantum teleportation and oth.) [19]. The problems of causality aberrations elevate in connection with the object representation of Universum. Spatial-temporally, it is impossible to explain the wholeness of the world of which experiments with the so-called «entangled» states - where spatially scattered objects immediately «feel» each other at any distance - are directly suggestive.

Third, rough determinism inevitably follows from the «primacy» of object Being. The world becomes a mechanistic closed system, as the subject and the categories connected with it - creativity, development, evolution and consciousness - are placed behind «the brackets» of physical Reality. The impossibility of evolution's full-value inclusion in object description leads to the problem of time arrow and reduction of wave function in quantum physics. Rough determinism and initial mechanisticity of Reality also excludes the occurrence of subject proper in it, and the problem of Consciousness becomes mystical.

In object description, there are problems with the description of the present as a moment (dot) and temporal (interval) measurement of any property (*e.g.* frequency) inevitably including the past of which Blessed Augustine wrote in his «Confession» [34]. Actually, it means the absence of some dynamics in the present and, hence, in the past and future (Zenon pradoxes and oth.). Herewith, there are cosmological paradoxes of the outset of Universe (Great Explosion) and, hence, the time proper which cannot enter the equations describing these phenomena.

In natural sciences, objectivity based on the total exclusion of the subject out of the explanatory paradigm leads to a logical deadlock, as the objective truth - under such an approach - is in the transcendental world on that side of the feeelings of the researcher. The list of problems may be continued.

It is necessary to note that any of our Reality representation, as a theoretical description of some phenomena (simulation), is possible only when it is presented in a closed system and is described by a finite content, and may be predicted for any moment of time. Our feelings and worries are always extended in time. Time is something «flowing» for us, but it belongs to the contentive formalisation of any process $f(t)$ as a parameter, and time does not flow anywhere! In this case, the temporal process may be opposed by whole stationary state of a system extended in time, but it does not depend on it. It becomes obvious if we pass on from the 4-dimensional spatial-temporal representation to the infinite Hilbert space and present $f(t)$ as a vector in the space of states. For instance, a signal $f(t)$ - using the Fourier transformation - may be presented as spectral $F(v)$. The occurring ratio of indefiniteness among presentations (in physics or Fourier window transformation in mathematics) is conditioned by the thing that the completeness of transformation between these spaces is possible only for infinite time. In principle, mathematically, these are equivalent descriptions. However, the point of which Reality representation is «primary» (what is the «source» of our feelings) and which is derivative (as Reality reflection in our consciousness) allows us to put everything in its place. First the answer seems obvious, just as the thing that Sun orbits the Earth, but not *vice versa*.

However, if the space of states is thought to be the «source» of our feelings (*wholistic subject* Reality representation unlike the one of the *object differentiated*), the above-denoted problems just disappear. In this case, realisation is a transformation of some pointed out whole non-local state of reality into the form of local spatial-temporal process which has *always been really carried out with an approximation*. For

---

[55] *E.g.*: if omnipotent and all-mighty God can create a stone which He will not be able to lift up. See also paradoxes in Kant multitudes theory.

example, in case of Fourier window transformation, averaging proceeds by some spectral window [56] to which certain temporal window corresponds, through which we «look over» (experience) the whole state successively as a process. Temporal window and precision of description are connected with the ratio of indefiniteness, and the present is determined not by a dot, but some interval (temporal window). Each realisation act is now «physically» irreversible and leads to the reduction of wave function in Heisenberg's interpretation [57]. Each perception act is already not the «reflection» of a transcendental thing from the «on-the-other-side» world, but it is the transition of a concrete Reality state into a localised spatial-temporal form of a process (phenomenon) in full correspondence with classical or quantum mechanics. Individual limited consciousness is now a subsystem (reference system [58] of Universum's subject representation as a whole and it is a temporal window through which a special particular state is «seen over» (or extended, to be more precise) as a «subjective process» localised in space. One can use such a simplified analogy: an individual can discern and «tune» to the state of separate organs (subsystems) and realise (experience) them as a dynamic localised process (feeling) from the wholistic state of all the organism. Herewith, the state proper and its transformation during its realisation are real «phenomena» and, thus, they condition each other during realisation. «Window transformation» originates indefiniteness which requires a state's objective reduction under localisation. We may treat the wholistic Reality representation as subject being, in principle, available to any «individual» as its special particular «reference system» (within this system's restrictions) and, thus, the process of realisation (reduction) will not depend on which reference system it proceeds in (that undoes the paradox of «Wigner's friend» [13]). Reduction is carried out not at the object (local), but at the systemic whole level - immediately and for all possible reference systems.

Spatial-temporal representation creates an illusion of object Being and time flow (course). But, when a painter creates a picture or a conversation goes, then we refer *the whole unfinished process* to our *present*. Does time exist as a moment? St. Augustine denied this. True, to perceive a sonic tone, it is necessary to determine its frequency. But, to determine it, a span of time is *physically* required, as minimum, equal to the period of at least one (here sonic) fluctuation. To set one harmonic physically, the whole spatial-temporal *continuum* is needed. So, does time «flow» anywhere beyond our consciousness? Neurophysiological investigations show that when receptive retina fields, reflected into external geniculate body, begin to pulsate, collapsing into a dot, we «go blind» for a while. Thus, motion continuity in our consciousness is an illusion created by the function of specific brain sections. In particular, people with a stroke-affected V5 field see the world as a sequence of motionless still frames. Actually, we perceive the world «in frames», and perception continuity is an illusion. Maybe, the Einstein Universe is just a frame in the perception of Nature from one creative act to the other? In the Poincare contribution of 1912 «Quantum hypothesis» [68], the statement about discreteness of a multitude of any isolated physical system's possible states - as Poincare asserts - is also applicable to Universe: «Hence, the Universe is to transfer from one state to the other in a «hop», but it remains unchanged in the intervals between hops, and different moments, during which it preserves its state, could not be differentiated from each other; thus, we come to the interruptive time flow, to atoms of time». Note that time is evolutionary, and discrete in the space of states - unlike the physical «continuous». The first one sets the order of states change, the second - their duration.

Absence of demarcation between perceptual and functional spaces and times leads to misfits. And all this takes place despite the thing that distinct demarcations, *e.g.* between real and perceptual spaces, were made in contributions of Bertrand Russell. In particular, he wrote: «Not only colours, sounds, *etc.* are absent in the world of scientific matter, but also space which we perceive with vision and touching. It is relevant for science that its matter is in space, but its space cannot be precisely the one we see and touch...real form science is focused on is to be in real space which should differ from the seeming space of every man [41]».

The point why this very form of consciousness turned to be evolutionarily profitable to us is quite interesting, but it is beyond our consideration. However, it is possible, at least in part, to analyse how Reality would be

---

[56] By the by, the presence of spectral window side-petals allows us to understand the effects of sub-barrier leakage of wave function.

[57] According to Heisenberg [4] « *quantum hop*», in observation, refers to a change of our knowledge.

[58] Countdown system is a physical analogy of the reference system; individual is biological, individuum is psychological, *etc.*

perceived in the space of states and is it possible to fully verbalise this experience? First, we enumerate the traits of such experience based on the theoretical representation of phenomena in the Hilbert space.

1. Experience of the world integrity without its division into objects (including body, individuality, *etc.*) and, as a consequence - absence of ego-experience and ego *per se*.

2. Absence of physical space and time, non-locality, including the absence of local feelings. Absence of restrictions (limitations) for velocities of changing any properties - actually, absence of processes extended in «physical time», including those psychical, mental and verbal processes. Absence of «processual rigidness» is experienced as moment immediateness under any changes. Qualitative changes prevail over the quantitative.

3. Experience of intuitive understanding of the world as a consequence of identity with it, but not by logical speculations.

4. Reality is experienced as qualitative, but not quantitative diversity. Gradations of quality intensity disappear.

Obviously, it is impossible to comprise these two types of consciousness and, «from the inside» of everybody, the other is *nothing (not realised)*. Let us underline that here it goes not about some changed states of common consciousness, but about its principally different types realised in completely different spaces. Is this type of consciousness achievable?

Having dropped mysticism, let us try to compare the consciousness of «space of states» to those experiences which occur during meditation (dhyana) and they correlate with the thing called Satoti Zen, Awakening (enlightment), Nirvana, Samadhi and oth. in the East [69].

The Chan principle reads: « do not base upon words and scriptures», and «do not stick to name and form». According to common acknowledgement of different Mahayana schools and trends, true Reality cannot be expressed in any linguistic means, the idea of non-verbal mastering the truth is connected with such categories of Budda philosophy as nirvana, shunyata (void), anatman (absence of individual «I») and others [70]. It is directly connected with the absence of any local process including also verbal ones in the space of states.

The proper meditation process is usually anticipated by prepartory exercises that alleviate physical tension and other negative factors being in the way of concentration. Meditation was realised by means of concentration of consciousness devoid of any images or thoughts in one point (Sanscr. - *ekagra*; Chin. *i-nyan-sin*) which agreed with conscious concentration of attention and stabilisation of consciousness. Usually, meditation began with conscious concentration of attention at one point (watching the *void*) and «devastation of consciousness» on the other. Such a state was called «single-dot» (Chin. *u-nyan-sin*) or «non-consciousness» (*u-sin*). This state was also called u-vo - «not me» as the world division into «I» and «not I» was absent in it. «It is the state of integrity», as A. Watts wrote [70]. The transfer from the object representation to that of subject, which is «nothing» related to the first one is completely adequate. Stopping of processual experiences (both physiological and mental) is an indispensable stage of changing the type of consciousness (see above).

It is worth noting that, in Buddism, proper meditation cannot be the cause of satori as it is beyond its «limits» (and it is shunyate - nothing) and it only increases the probability of spontaneous switch-over of consciousness type. Moreover, wish to acquire Nirvana only enhances ego-concept and levels it down to the thing that exists only in the object space-time. The wish of proper nirvana is already consciousness limitation, and it becomes a brake on the way of «awakening». As Zen reads: you cannot acquire true pureness destroying impureness as these are interconnected concepts, just as it is impossible to destroy all lowlands and, at the same time, expect that all highlands will remain [71]. Satori was achieved spontaneously and this process was described metaphorically - «keg's bottom suddenly falls off», - it is quite a figurative description of the hop from the finite and limited into the infinite and unlimited.

Analogous statements can be also found in Daosism. Chzhuan Tszi wrote: «As if my body fell off me, and reason faltered. As if I left my burden cover, rejected knowledge and imitated the all-pervading [72] «. «I look at it and cannot see it; so I call it invisible. I listen to it and cannot here it; so I call it inaudible. I try to seize it, but do not reach it; so I call it tiny. You need not aim to know about the source of it because it is uniform. It is infinite and cannot be named [72]».

Note that Sansara (object world) in Mahayan and Nirvana in Buddism are identical in essence - these are not two realities, but the *two types of its representation*. Sansara cognition is an infinite process of successive approach to the truth, but it is social, as knowledge transfer is possible semiotically. Nirvana cognition is a momentary act of intuitive understanding, but totally individual and «out-of-personality» cognition way. Its adequate translation is impossible in the transfer to an individual localised and limited form of consciousness in Sansara.

As we see, the coincidence is full. It allows us to treat the practice and psychotechnics - having been developed for more than 2000 years within present traditions and not to think them to be their artefact of mystical and ecstatic states of consciousness or «extension» of common consciousness (or, as a variant, a changed state of consciousness) - in quite a serious way. It is principally the other form and type of consciousness. It is also important that the classical scientific (object-oriented) method of investigation is not adequate here. Herewith, a systemic analysis would be more suitable with its teleological principles [73]. However, such an approach was only declared by its founders [73, 74] and, essentially, it has not been realised in practiced. A new paradigm's reduction to the common object form, in which the announced concepts of purpose, evolution and subject just do not find their right places, was one of the reasons for such a state.

With the present conclusion, we would like to preface the transfer to the next volume devoted to the problems of Consciousness.

APPENDIX A

To understand the principle of coding, we will have to begin with prime numbers. Prime number is the one which cannot be decomposed into co-factors, but, surely, into 1 [75]. Prime numbers are of exclusively big, both scientific and applied, importance, especially in cryptography. First, let us determine the operation of «to be congruent modulocomparison on module». Considering remainders resulting in division of different numbers into some other number, *e.g.* 6 will obtain:

$$
\begin{array}{lll}
1 = 0 \times 6 + 1 & 7 = 1 \times 6 + 1 & -5 = -1 \times 6 + 1 \\
2 = 0 \times 6 + 2 & 8 = 1 \times 6 + 2 & -4 = -1 \times 6 + 2 \\
3 = 0 \times 6 + 3 & 9 = 1 \times 6 + 3 & -3 = -1 \times 6 + 3 \\
4 = 0 \times 6 + 4 & 10 = 1 \times 6 + 4 & -2 = -1 \times 6 + 4 \\
5 = 0 \times 6 + 5 & 11 = 1 \times 6 + 5 & -1 = -1 \times 6 + 5 \\
6 = 1 \times 6 + 0 & 12 = 2 \times 6 + 0 & 0 = 0 \times 6 + 0
\end{array}
$$

It is clear that only one of the numbers that may be 0, 1, 2, 3, 4, 5 can be a remainder under division into 6. Numbers a and b are said to be comparable on module 6 if the remainder is the same in division into 6. In the above-enumerated sequence of numbers, those 5, 11 and -1 are comparable on module 6. In other words, ratio $a - b = nd$ - where d - divider, n - integer - should be fulfilled. For the ratio of comparison, a special indication is introduced: two integers a and b are said to be congruent to modulo d, then it is written down as: $a \equiv b(\bmod d)$ (do not confuse it with «identity» sign).

Comparisons turn to have many properties of usual equations. For instance:

it is always so that $a \equiv a(\bmod d)$;

if $a \equiv b(\bmod d)$, then $b \equiv a(\bmod d)$;

if $a \equiv b(\bmod d)$ and $b \equiv c(\bmod d)$, then $a \equiv c(\bmod d)$;

if $a \equiv a'(\bmod d)$ and $b \equiv b'(\bmod d)$, then $\begin{cases} a+b \equiv a'+b'(\bmod d) \\ a-b \equiv a'-b'(\bmod d) \\ ab \equiv a'b'(\bmod d) \end{cases}$

Modular arithmetic on one and the same *modulus* can be summed, subtracted and multiplied. Comparisons can be represented as «rolling-on» of numeric axis on the circumference (Fig. (**3.31**)).

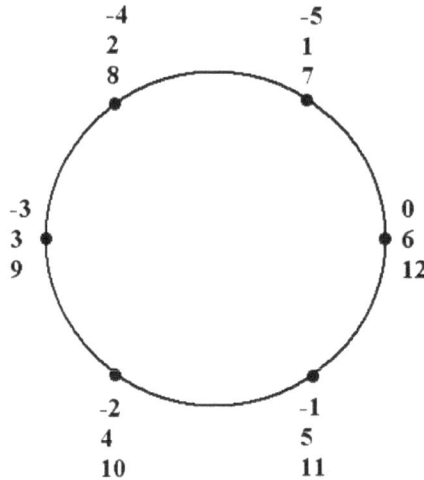

Figure 3.31. Congruence modulo 6.

In the 17th century, Pierre de Fermat proved the following theorem. If p is a prime number, not dividing integer a, then $a^{p-1} \equiv 1(\bmod p)$, i.e. $(p-1)$-power of a - divided by p - results in remainder 1. For example $10^6 \equiv 1(\bmod 7)$, $2^{12} \equiv 1(\bmod 13)$ and so on. The proof of the theorem can be found in the book [76]. Euclid algorithm and Eiler function are used in modern coding algorithms. Euclid algorithm is derived from the simple procedure of division of one number by the other. For two numbers a and b, let there be common dividers among which number d is the greatest common divisor. This number is precisely called as *the greatest common denominator, greatest common factor*, or *highest common factor*, and is indicated as $d = (a,b)$. Euclid algorithm is based on the ratio derived from the operation of division of one number by the other: $a = b \cdot q + r$; hence it follows that $(a,b) = (b \cdot r)$.

$$\begin{cases} a = bq_1 + r_1 & (0 < r_1 < b) \\ b = r_1 q_2 + r_2 & (0 < r_2 < r_1) \\ r_1 = r_2 q_3 + r_3 & (0 < r_3 < r_2) \\ r_2 = r_3 q_4 + r_4 & (0 < r_4 < r_3) \end{cases}$$

In this scheme, the greatest common denominator is equal to the last remainder r_n in the above-mentioned sequence. This algorithm is used in the solution of linear diophant equation $a \cdot x - b \cdot y = 1$ at $(a,b) = 1$ and $a > 0$, $b > 0$.

It turns out that a number's computer-assisted simplicity identification is very fast using modern methods, and decomposition of big numbers into factors is an impractical task for modern ECM for a reasonable time. It is this thing that modern coding systems are based on.

Now let us determine Eiler function $\varphi(n)$. Numbers a and b are said to be co-prime or relatively prime if their greatest common denominator is equal to 1, *i.e.* $(a,b) = 1$. Then, for n - an arbitrary positive number - let us indicate the number of such integers within the limits from 1 to n, which is mutually simple with number n. To make it vivid, let us write out $\varphi(n)$ values for the beginning of the numeric set.

As is seen from the presented sequence, the value of Eiler function $\varphi(p) = p - 1$ is for any prime number p, as such a number has no other dividers but 1 and p. Thus, Fermat theorem can be formulated as $a^{\varphi(p)} \equiv 1 (\text{mod } p)$.

$$\begin{cases} \varphi(1) = 1 \Rightarrow 1 \\ \varphi(2) = 1 \Rightarrow 1 \\ \varphi(3) = 2 \Rightarrow 1,2 \\ \varphi(4) = 2 \Rightarrow 1,3 \\ \varphi(5) = 4 \Rightarrow 1,2,3,4 \\ \varphi(6) = 2 \Rightarrow 1,5 \\ \varphi(7) = 6 \Rightarrow 1,2,3,4,5,6 \\ \varphi(8) = 4 \Rightarrow 1,3,5,7 \\ \varphi(9) = 6 \Rightarrow 1,2,4,5,7,8 \\ \varphi(10) = 4 \Rightarrow 1,3,7,9 \end{cases}$$

Let Alice and Bob decided to begin their secret correspondence, for which they chose prime number p. Then each of them chose, reporting to nobody, a natural number - coprime with number $p-1$: Alice - $a1$ and Bob - $b1$. Then each of them found numbers $a2$ and $b2$, such that:

$$a1 \times a2 \equiv 1 \left(\text{mod } \varphi(p) \right), \quad 0 < a2 < p - 1 \text{ and}$$

$$b1 \times b2 \equiv 1 \left(\text{mod } \varphi(p) \right), \quad 0 < b2 < p - 1.$$

These two numbers are the secret keys of Alice and Bob. Let us analyse a simple example in which Alice wants to transmit number $0 < x < p - 1$. To begin with, she codes it with her first key $a1$ the following way:

$x1 \equiv x^{a1} (\text{mod } p), 0 < x1 < p$ and transmits the result to Bob. He, in his turn, carries out the operation $x2 \equiv x1^{b1} (\text{mod } p), 0 < x2 < p$ using his first key and transmits the received value back to Alice. She codes it with her second key.

$x3 \equiv x2^{a2} (\text{mod } p), 0 < x3 < p$ and sends it to Bob who carries out the last operation with his second key $x4 \equiv x3^{b2} (\text{mod } p), 0 < x4 < p$, the obtained value being $x4 = x$, which follows from:

$$x4 \equiv x^k (\text{mod } p), k = a1 \times a2 \times b1 \times b2 (\text{mod } p - 1)$$

$a1 \times a2 \equiv 1 \left(\text{mod } \varphi(p) \right), b1 \times b2 \equiv 1 \left(\text{mod } \varphi(p) \right)$ and hence, $k \equiv 1 \left(\text{mod } \varphi(p) \right)$ and $x4 = x$.

We demonstrate it with a small prime number $p = 23$ (open key), and let Alice choose $a1 = 5$ and Bob - $b1 = 7$. Then the solution of $5 \cdot a2 \equiv 1 \left(\text{mod } \varphi(23) \right)$ will be $a2 = 9$ and that of $7 \cdot b2 \equiv 1 \left(\text{mod } \varphi(23) \right)$ will be - $b2 = 19$.

Alice decides to transmit the code of a combination lock, *e.g.* Fig. **17**; then she sends the division remainder $17^5 = 61732 \times 23 + 21$ or $17^5 \equiv 21 (\mathrm{mod}\, 23)$, *i.e.* 21.

With his first key, Bob recodes the code he obtained and gets $21^7 = 78308197 \times 23 + 10$ or $21^7 \equiv 10 (\mathrm{mod}\, 23)$, *i.e.* 10, which he sends back.

Alice uses her second key and sends the result $10^9 = 43478260 \times 23 + 20$ or $10^9 \equiv 20 (\mathrm{mod}\, 23)$, *i.e.* 20.

And, finally, Bob receives $20^{19} + 22795130434782608695 6521 \times 23 + 17$ or 20^{19}

$\equiv 17 (\mathrm{mod}\, 23)$, *i.e.* 17 - the number Alice wanted to transmit.

APPENDIX B

Earlier we already used the term «quasi-time» regarding the description of quantum system. Let us analyse its semantic aspect in other physical applications and consider the situation in which two individuals perceive a process determined by a sequence of the changing parameter x: $\{x_1, x_2, ..., x_T, ...\}$, characterised by a certain intensity of V_j property. In physics, the simplest parameter of V_j property of a body's velocity of even and linear movement. In this case, parameter x is the body-covered way for the time $\{t_1, t_2, ..., t_T, ...\}$.

Naturally, all V_{j_1} values, determining the first individual, in its «own» reference system can be made equal to zero (for it, they are initial points setting its «co-ordinate system»), and the second (V_{j_2}) can rendere the intensity value of the considered property equal to $\Delta V_j = V_{j_2} - V_{j_1}$.

Now we introduce the notion of «absolute» time T as an index [59] which is «rendered» by the subject in its «own» reference system to all successive forms of processes proceeding in the world:

$$x_1 \xrightarrow{\hat{S}} x_2 \xrightarrow{\hat{S}} \cdots \xrightarrow{\hat{S}} x_T$$

Here S is the operator that determines the sequence of forms (values) of some process. As dynamicity (velocity of changing x) should depend on the rigidness or inertia of a concrete quality, we can try to relate its variability to V_{H_j} because this index is connected with a concrete quality (j) and it is obtained from that of U_H characterising an object's rigidness.

Now we determine $V_{H_{j_1}}$ as an «indexing velocity» of parameter x («time flow velocity» in subjective perception) or rigidness of j-property from the first reference system. Subjective time of realisation (from index 0 to T) of the considered process will be determined by the first individual as $t = V_{H_1} \cdot T$, and $t' = V_{H_2} \cdot T$ - by the second. It will mean that dynamicities (or flow velocities) of processes are perceived in a different way in various reference systems.

Let us check our supposition. Having used transformation (formula 3.15), we will obtain:

---

[59] Of the index enumerating the sequence of forms $x_1, x_2, ..., x_T$ of a proceeding process.

$$t' = T \cdot V_{H_{j_2}} = \frac{T \cdot V_{H_{j_1}} \cdot \sqrt{1 - \frac{\left(\Delta V_j\right)^2}{C^2}}}{1 + \frac{V_j \cdot \Delta V_j}{C^2}} = \frac{t \cdot \sqrt{1 - \frac{\left(\Delta V_j\right)^2}{C^2}}}{1 + \frac{V_j \cdot \Delta V_j}{C^2}}$$

Hence,

$$t = \frac{t' + \frac{\Delta V_j}{C^2} V_j \cdot t'}{\sqrt{1 - \frac{\left(\Delta V_j\right)^2}{C^2}}} \tag{3.38}$$

If property V is velocity, then $V \cdot t' = x'$ and

$$t = \frac{t' + \frac{\Delta V}{C^2} x'}{\sqrt{1 - \frac{\left(\Delta V\right)^2}{C^2}}} \tag{3.39}$$

determine the temporal transformation for physical processes in the transition into another inertial countdown system which precisely coincides with the well-known Lorentz transformation and proves our supposition about V_H as the characteristics that determines subjective velocity of time flow. Note that «time deceleration» effects can be observed not only in physics in case of extremely big velocities, but also in psychology under extreme changes of an individual's [60] psychical traits. Many people personally experienced the change of time flow tempo in stressful situations.

It is possible to continue the proof of this interpretation in the accordance with different physical phenomena. For instance, we will obtain the equation of Doppler effect for a wave process with τ effect having

$$\tau' = V'_H \cdot \Delta T = \frac{\Delta T \cdot V_H \cdot \sqrt{1 - \frac{\left(\Delta V\right)^2}{C^2}}}{1 + \frac{\Delta \vec{V} \cdot \vec{C}}{C^2}} = \frac{\tau \cdot \sqrt{1 - \frac{\left(\Delta V\right)^2}{C^2}}}{1 + \frac{\Delta V}{C} \cos \theta}$$

As $\tau = 1 / \omega$, then frequency ω' is equal to:

$$\omega' = \frac{\omega \cdot \sqrt{1 - \frac{\left(\Delta V\right)^2}{C^2}}}{1 \pm \frac{\Delta V}{C} \cos \theta} \tag{3.40}$$

---

[60] In psychiatry, the so-called «hypermnesia of the drowning» (official name), which is expressed in death-endangered people, is described. People that underlived the inevitable assert the thing that all their live pictures passed by before them in every small detail, beginning with the last moments and ending in the images of their childhood. So, Francis Beaufort, an English hydrographer and admiral that was saved after the shipwreck, experienced hypermnesia and described it in his times. A change of temporal scale (both towards acceleration and deceleration) can also be set using hypnotic suggestion [77]. Deceleration of the tempo of biological aging is also known in sopor. Many people personally experienced the change of time flow tempo in stressful situations.

From ratios 3.14 and $\dfrac{E^2}{C^2} = P^2 + m_o^2 C^2$, we can get the Lorentz co-ordinate transformation:

$$x' = V' \cdot t' = \frac{V + \Delta V}{1 + \dfrac{V \cdot \Delta V}{C^2}} \cdot \frac{t + \dfrac{\Delta V}{C^2} x}{\sqrt{1 - \dfrac{(\Delta V)^2}{C^2}}} =$$

$$= \frac{\dfrac{\overset{\frac{x}{V \cdot t}}{V \cdot t} + \dfrac{V \cdot \Delta V}{C^2} \cdot x + \Delta V \cdot t + \dfrac{(\Delta V)^2}{C^2} x}{\left(1 + \dfrac{V \cdot \Delta V}{C^2}\right) \cdot \sqrt{1 - \dfrac{(\Delta V)^2}{C^2}}} =$$

$$= \frac{x \cdot \left(1 + \dfrac{V \cdot \Delta V}{C^2}\right) + \Delta V \cdot t \left(1 + \dfrac{\Delta V}{C^2} \cdot \overset{\frac{V}{x}}{\dfrac{x}{t}}\right)}{\left(1 + \dfrac{V \cdot \Delta V}{C^2}\right) \cdot \sqrt{1 - \dfrac{(\Delta V)^2}{C^2}}} =$$

$$= \frac{\left(1 + \dfrac{V \cdot \Delta V}{C^2}\right) \cdot (x + \Delta V \cdot t)}{\left(1 + \dfrac{V \cdot \Delta V}{C^2}\right) \cdot \sqrt{1 - \dfrac{(\Delta V)^2}{C^2}}}$$

We obtain

$$x' = \frac{x + \Delta V \cdot t}{\sqrt{1 - \dfrac{(\Delta V)^2}{C^2}}} \tag{3.41}$$

Hence, the transformation rules of velocity (3.9), time (3.39), co-ordinates (3.41), Doppler effect (3.40) are, essentially, a consequence of a certain way of Reality representation in our mentality which, being beyond the conditions of natural biological adaptation, close to «ideal» C (at $\varphi \approx 0$) leads to its paradoxical perception, all relativistic effects being, in essence, *not physical, but psychological* (mental).

Note that, in the semantic space, it is possible to compare anti-particle $\vec{u}^{(-)} = \{-u_1, -u_2, ..., -u_n, -u_H\}$, having an inversion over all properties and time [61] (as «time arrow» is determined by sign u_H, as $t = V_H \cdot T$), to the description of any particle $\vec{u}^{(+)} = \{u_1, u_2, ..., u_n, u_H\}$. Actually, temporal inversion means a reverse change in the direction of a particle's motion, which is equivalent to *the change of charge sign* determining the force action gradient. Besides, if we accept the theory of GREAT EXPLOSION and the origin of the Universe out of vacuum, then the sum of all semantic co-ordinates should be equal to zero in it. It is possible only in case if the presence of anti-world with the negative u_H, which «drifted apart» from our Universe, is presupposed in time at the very outset of its origin [62].

Thus, we showed that, in any theory, an object's adequate description through its properties is possible only in the semantic space and requires an obligatory indication of not only properties intensities, but also their

---

[61] In physics, CPT theorem shows the symmetry of the laws of nature with simultaneous *spatial inversion* (P), *time circulation* (T) and *charge conjugation* - particles replacement for corresponding anti-particles.

[62] Which explains why the Universe did not annihilate right after its origin.

rigidness. It is connected with the thing that properties that are presented not in one-dimensional $(\vec{e}_j)$, but plane *continuum* $(\vec{e}_j \times \vec{e}_H)$, as it was shown above.

Therefore, the psychosemantic way of objects description (objects representation on the mental map) we chose leads to a complete accordance with physical Reality outlook. Absolutely, the obtained ratios are *objective and general* for mental maps of all *Homo sapiens* individuals. However, they are related *not to Reality proper, but to the mental way of its representation.*

In psychology, it is possible to check relativistic relations using psychophysical material, as this is the only field where psychologists make quite precise measurements. An extended algorithm of spatial construction is presented in **Appendix C**.

APPENDIX C

An example of using relativistic psychometrics in psychophysics

Obviously, it is practically impossible [63] to provide an acceptable level of signals differentiation in one range for the majority of sensor systems. The phenomenon of sensor apparatus adaptation to average stimulus, known in psychophysics, is connected with it. For instance, going out of a dim room to the bright sunlight, we «go blind» for a while until our sensor system «switches over» to a new range. Eye sensitivity is known to increase 200 000 times [64] [78] in the transfer from bright luminosity into darkness.

Let us demonstrate the practical application of relativistic scales in psychology with the example of interconnection between subjective loudness and sound intensity. First, we remember the basic psychophysical statements in this field and then pass on to the calculations on our methods.

Sonic stimulus intensity (L) is accepted to be determined in decibels (dB): $L = 20 \cdot \lg\left(\dfrac{P}{P_0}\right)$, where $P_0 = 2 \cdot 10^{-5}\ Pa$.

Signal intensity (L) of 40 dB at frequency $f = 1000$ Hz is standardized in physics as subjective loudness (N) of 1 sone or loudness level (L_r) of 40 phons (at sound frequency of 1000 Hz, sonic pressure scale in dB`s and the scale of loudness level in phons coincide).

In psychophysics, measurements of loudness curve begin with the initial level of 40 dB (1 sone) in relation to which the level of sound intensity, subjectively percepted as lowder twice, is determined. This is the way a set of sound intensity values - each of which corresponds to double loudness related to the previous one, *i.e.* 2, 4, 8, 16, 32, 64 and 128 sones, respectively - is found. Then this series of experiments is arranged in the reverse way from the base value (40 dB), and a number of sound intensity values - each of which is twice less than the previous one: 1/2, 1/4, 1/8, 1/16, 1/32 sone [79] - is determined. According to the obtained experimental points, the corresponding curve of the interconnection between physical value (sound pressure level) and loudness, which is the index of sensation, is made up. This curve is shown in Fig. (**3.32**). Similar regularities between factors of stimulus and perception were found by S. Stevens for many sensations of different modalities. In the right part of Fig. (**3.32**), several diagrams of S. Stevens power function (equality 3.42) - by which these curves are described - are shown.

---

[63]*E.g.* it is 130 dB on loudness scale.

[64]It happens due to a change of eye pupil's diameter and pigment motion in the retina.

$$N = const \cdot \left(\frac{P}{P_0}\right)^{2k}$$

(3.42)

In the region of levels higher than 40 dB, this connection is approximately reflected with a straight line (on logarithmic scale). Therefore, an international agreement was set on the thing that, in the range of 40 - 100 phon, the interconnection between loudness N (sones) and level L_r should be expressed with the following formula:

$$N = 2^{\left(\frac{L_r-40}{10}\right)}.$$

(3.43)

The schedule of the function set by expression 3.43 is shown by the dashed line. It is seen that this straight line reconciles well with the experimental one only at the levels higher than 40 phons.

Now, using formula (3.13), let us theoretically calculate the interconnection between subjective loudness and sound intensity. For average stimulus intensity of L_i dB and frequency 1000 Hz, the ratio will be:

$$\frac{P_0 \cdot 10^{\frac{L}{20}}}{P_0 \cdot 10^{\frac{L_i}{20}}} = \frac{\dfrac{U_H \cdot N}{\sqrt{1-\dfrac{N^2}{C^2}}}}{\dfrac{U_H \cdot N_i}{\sqrt{1-\dfrac{N_i^2}{C^2}}}} = \frac{N \cdot \sqrt{C^2 - N_i^2}}{N_0 \cdot \sqrt{C^2 - N^2}}$$

(3.44)

where N_i - base (average in the range) loudness level corresponding to the stimulus intensity of L_i dB, C - loudness limits value in the i -range.

According to the scheme of experiment, we suppose that L_0 = 40 dB (i = 0) determines L at which subjective loudness becomes twice higher. Stimulus intensity is calculated in the formula:

$$\frac{P_0 \cdot 10^{\frac{L}{20}}}{P_0 \cdot 10^{\frac{L_i}{20}}} = \frac{2 \cdot \sqrt{C^2 - 1^2}}{1 \cdot \sqrt{C^2 - 2^2}}$$

(3.45)

As first double-loudness occurs at L = 50 dB (2 sones), then, therefore, we can calculate: $C = \sqrt{6} \approx 2.449$.

Substituting L_1 = 50 to the left part of equality (3.45) and accepting the subjective loudness value that corresponds to it as unit (new subjective loudness reference value), we will calculate the second doubling of loudness (4 sones).

At the calculated value L = 60 dB, sound loudness is again accepted as unit and, substituting L_2 = 60 dB to the left part of the equality (3.45), we will calculate the next value L = 70 dB of loudness doubling (8 sones). In repitition of this operation, we successively get the following values: 80 dB - 16 sones, 90 dB - 32 sones, 100 dB - 64 sones, 100 dB - 128 sones, 120 dB - 256 sones. Thus, we *theoretically* obtained the values which precisely correspond to *experimental* investigations and S. Stevens formula.

The diagram of loudness Fig. (**3.33**) has a bend precisely in point 40 dB accepted as loudness unit by Stevens. The thing is not in the exclusive intuition of Stevens when it was 50 dB that he chose. If Steves had accepted the subjective loudness value of 50 dB as 1 sone, then the diagram's bend would have been likely in this point. It is connected with the thing that relativistic scales *are not symmetrical* regarding the increase or decrease of sensations related to the reference value. Indeed, we should use the following equality at a successive double-decrease of loudness:

$$\frac{P_0 \cdot 10^{\frac{L}{20}}}{P_0 \cdot 10^{\frac{L_i}{20}}} = \frac{\frac{1}{2} \cdot \sqrt{1 - \frac{1^2}{C^2}}}{1 \cdot \sqrt{1 - \frac{(1/2)^2}{C^2}}} = \frac{\sqrt{C^2 - 1}}{\sqrt{4 \cdot C^2 - 1}} \tag{3.46}$$

It was rather complicated to compare the theoretical values to experimental data in the lower part of the curve. In the paper [80], the authors present the diagram they obtained in the region lower than 40 dB (see Fig. (**3.33**)).

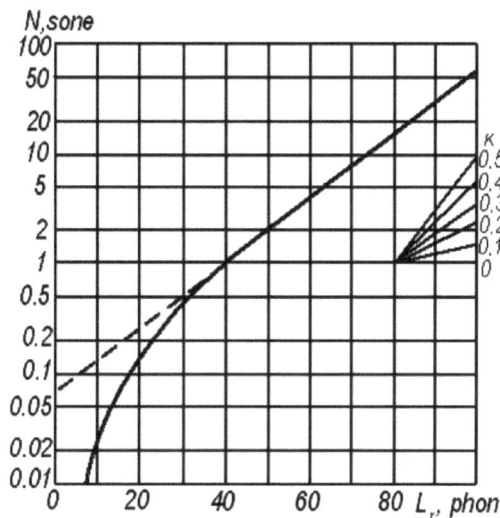

Figure 3.32. Interconnection between subjective N and level of loudness $_rL$ [79].

In the diagram, the authors also presented the experimental points obtained by other researchers. When making measurements, many researchers combine both the method of «ascension» (double on-growth of subjective loudness) and «descent» (double decrease of subjective loudness) thinking them to be equivalent regarding the obtained data [79] [65]. Therefore, the character of experimental points location in Fig. (**3.33**) is often just unfathomable based on experimental logic. For instance, it is not clear how - with the experimental methods used - researchers obtained their subjective loudness values higher or lower than 1/2, 1/4, 1/8, 1/16, 1/32 sone. It is also not clear how could it turn out that, at practically the same sonic pressure value of 31 and 32 dB, Scharf and Stevens [81] got the points of subjective loudness equal to 0.28 and 0.47 sone, respectively (see Fig. (**3.33**))? This list of questions could be continued.

Using our methods (descent method - formula (3.46), we precisely calculated the points at the successive decrease of signal intensity beginning with 40 dB. We showed the obtained points in Fig. (**3.33**) (they are marked with letter **A**). It is seen that our points quite well accord with the general tendency of the obtained experimental data, being precisely adequate to the measurement methods, *i.e.* they have values of 1/2, 1/4, 1/8, 1/16, 1/32 sone.

Besides, using our calculation methods, one can try to explain the presence of groups of experimental points at ≈ 10, ≈ 30, ≈ 36 dB, which do not «come out» when using descent method (see Fig. (**3.34**)). For

---

[65] It is not really so (see equalities 3.45 and 3.46). Moreover, researchers often calculate average values on these methods.

this purpose, we use the ascension method (expression 3.46) and the calculated **A** points values (obtained with the «ascension» method (see Fig. (**3.32**)) at $\approx$ 16, $\approx$ 27, $\approx$ 34 dB, respectively. Note that **B** points calculated with the ascention method and initial (reference) **A** points have one and the same subjective loudness values in sones. It is correct as reference values of subjective loudness, *i.e.* subjective perception of certain sonic pressure levels, are the condition for obtaining the points. It is also noteworthy that the calculated **B** points practically coincided with the groups of indicated experimental points. It indirectly proves the fact that, to obtain experimental data, researchers used descent and ascension methods, thinking them to be equivalent which is really wrong.

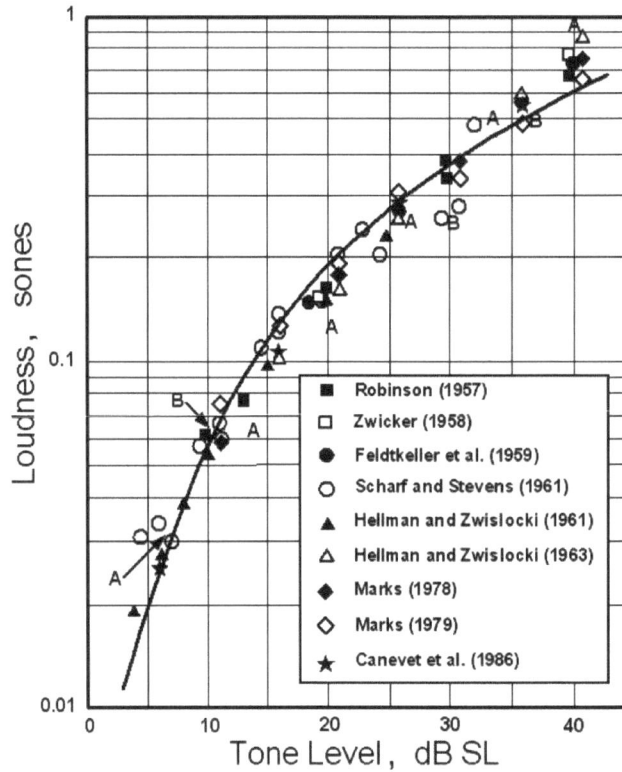

Figure 3.33. Loudness curve [75].

One should note that formula (3.44) can be correct related to lower sensation thresholds in the following way:

$$\frac{P_0 \cdot 10^{\frac{L}{20}} - P_0 \cdot 10^{\frac{L_{min}}{20}}}{P_0 \cdot 10^{\frac{L_i}{20}}} = \frac{\dfrac{U_H \cdot N}{\sqrt{1 - \dfrac{N^2}{C^2}}}}{\dfrac{U_H \cdot N_i}{\sqrt{1 - \dfrac{N_i^2}{C^2}}}} = \frac{N \cdot \sqrt{C^2 - N_i^2}}{N_0 \cdot \sqrt{C^2 - N^2}}$$

where L_{min} - threshold intensity value of sonic stimulus. Unfortunately, the lower sensation threshold is determined quite approximately and its value lies within the interval of confidence from 0 to 7 dB at average value equal to 3 dB [79].

Note that there is, really, little sense in the loudness curve as it just connects the various sites ranges which the comparison of loudness is made. As an illustration, using formula (3.45), we made up 4 ranges corresponding to signal levels L_0 = 40 dB, L_2 = 60 dB and L_4 = 80 dB, L_6 = 100 dB (see Fig. (**3.34**)) which are used in traditional experimental investigations. Continuous curve - 9 power polynomial regression calculated on theoretical loudness values obtained on formulae (3.45) and (3.46). Loudness curves for ranges corresponding to $L_0 = 40$ dB, $L_2 = 60$ dB, $L_4 = 80$ dB, $L_6 = 100$ dB are shown with dotted lines.

Let us explain the essence of obtained ranges for the traditional methods of loudness doubling. With the presence of the initial sonic field level of L dB, our ear gets tuned to the range from N_{min} to $(N \cdot C)$, where $N_{min} = 0$ sone (L_{min} - lower sensation threshold in *this range* - corresponds to it), and $(N \cdot C)$ - limit value of subjective loudness in sones for this range. For example, the initial sonic field level is 40 dB; then our ear gets tuned to the range from 0 sone to $(1 \cdot 2.449)$ sones, which corresponds to the values of sonic stimulus from $L_{min} \approx 3$ dB to ≈ 53 dB (see Fig. (**3.34**)), *etc.* Also note that L_{min}, corresponding to 0 sone, changes in the transition to the next range (see Fig. (**3.34**)). It is understandable, as the lesser average loudness level is, the weaker signals our ear can perceive and *vice versa*.

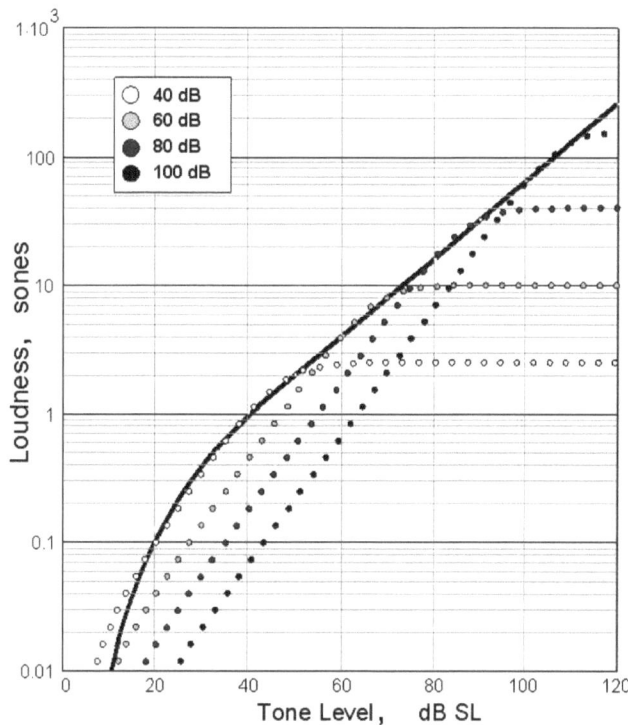

Figure 3.34: Loudness ranges.

Obviously, in adaptation, sensor apparatus is tuned to more informative frequency and dynamic ranges. One can check how the dynamic range changes if it is necessary to differentiate the signals not only in 2, but in 4, 8, 16, *etc.* times. For these cases, we calculated the ranges of C values (see Table **3.5**). It is seen from Table **3.3** that, if it is necessary to compare the loudness of 32 sones, compared to that of reference 1 sone, $C \rightarrow \infty$. It means that the maximum of dynamic range can not overlap 90 dB, *i.e.* the sensor apparatus cannot compare such differences in loudness level at *one* range.

Table 3.5: Calculated loudness ranges parameters.

Signal intensity (L), dB	Subjective loudness (N) sones	Limit loudness value in range (C), sones
50	2	2.449
60	4	4.344
70	8	8.265

| 80 | 16 | 16.203 |
| 90 | 32 | $> 10^{307}$ |

Note that now, using our formulae, a researcher has a possibility to make up dependences of loudnesses at arbitrary subjective loudness values and signal level.

Thus, we have shown that:

1. The formula we theoretically obtained *correctly describes the whole curve*, unlike experimentally obtained formulae that are functional in separate curve sites.

2. In subjective loudness measurement, descent and ascension methods *are not equivalent* and, thus, it is necessary to use relativistic scales.

3. In psychophysics, loudness curve is really not continuous, and it just *connects the sites of various ranges* within which the comparison of sonic loudness is realised.

REFERENCES

[1]　Y.I. Manin, *Computable and noncomputable (in Russian)*. Sov. radio, Moscow, 1980.

[2]　R.P. Feynman, "Simulating Physics with Computers," *Inter. Jour. Theor. Phys.,* vol. 21, no. 6/7, pp. 467-488, 1982.

[3]　J.F. Werker, R.N. Desjardins, "Listening to speech in the first year of life: Experiential influences on phoneme perception," *Curr. Direct. Psychol. Scien.,* vol.4, no.3, pp. 76-81, 1995.

[4]　V. Heisenberg, *Physics and Philosophy. Part and Whole (in Russian)*. Nauka, Moscow, 1989.

[5]　M. Born, *Ed., The Born-Einstein Letters*. Walker, NY., page 82, 1971.

[6]　Henry Head. In conjunction with W. H. R. Rivers, G. Holmes, J. Sherren, T. Thompson, G. Riddoch. *Studies in neurology*. Oxford Univ. Press, 2 vols. London, 1920.

[7]　S. Freud, *Intrtoduction to Psychoanalysis: Lectures (in Russian)*. Nauka, Moscow, 1989.

[8]　B.F. Skinner, *About Behaviorism, V.2*. Vintage, N.Y., 1976.

[9]　N. Bohr, *Selected Scientific Works (in Russian)*. Nauka, Moscow, 1971.

[10]　A.A. Reformatsky, *Intrtoduction to Linguistics (in Russian)*. Nauka, Moscow, 2002.

[11]　A.A. Markov, N.M. Nagorny, *Theory of Algorithms (in Russian)*. Nauka, Moscow, 1984.

[12]　J.V. von Neumann, *Mathematishe Grundlagen der Quantenmechanic*. Verlag von Julius Springer, Berlin, 1932.

[13]　E.P. Wigner. Remarks on the mind-body question. In *Quantum Theory and Measurement (Eds J A Wheeler, W H Zurek)* Originally published in The Scientist Speculates (Ed. L G Good) (London: Heinemann, 1961) Princeton: Princeton University Press, 1983.

[14]　H. Everett, *The Many-Worlds Interpretation of Quantum Mechanics*. Princeton University Press, Princeton, N.J., 1973.

[15]　B.S. DeWitt, N. Graham, *Quantum Theory and Measurement (Eds J A Wheeler, W H Zurek)*. Princeton University Press, Princeton, N.J., 1983.

[16]　R.P. Feynman, R.B. Leighton, M. Sands, *Feynman lectures on physics*. V. 3, Addison Wesley Publising Com., Inc., Reading, Massachusetts, Palo Alto, London, 1965.

[17]　L.V. Tarasov. *Bases of quantum mechanics (in Russian)*. Vysshaya shkola publ., Moscow, 1978.

[18]　E. Rieffel and W. Polak, "An Introduction to Quantum Computing for Non-Physicists," *ACM. Computing. Surveys.,*vol. 32, no. 3, pp. 300-335, 2000.

[19]　G. Greenstein, A.G. Zajonc, *The Quantum Challenge. Modern Research on the Foundations of Quantum Mechanics*. Jones and Bartlett Publishers, Inc., 2006.

[20]　R.L. Gregory, *The Intelligent Eye*. Wiedenfeid and Nicolson, London, 1970.

[21]　G. Somjen, *Sensory Coding in the mammalian nervous system*. Appleton-Ceture-Crofts Ed. Div. Meredith Corp., N.Y., 1972.

[22]　I.M. Kobozeva, *Linguistic Semantics (in Russian)*. Editorial URSS, Moscow, 2000.

[23]　H.H. Harman, *Modern Factor Analysis*. University of Chicago Press, Chicago, 1976.

[24]　P.A.M. Dirac, *The Principles of quantum mechanics*. At The Clarendon Press, Oxford, 1958.

[25] A.D. Logvinenko, *Measurement in Psychology: mathematical foundations (in Russian)*. Moscow University Press, Moscow, 1993.

[26] V.F. Petrenko, *Fundamentals Psychosemantics (in Russian)*. Moscow University Press, Moscow, 1997.

[27] A.R. Luria. *Language and Mind (in Russian)*. Feniks, Rostov-na Donu, 1998.

[28] B. Schumacher, "Quantum Coding," *Phys. Rev. A,* vol. 51, no. 4, pp. 2738-2747, 1995.

[29] K.A. Valiev, A.A. Kokin, *Quantum computers: hopes and reality (in Russian)*. Regular and chaotic dynamics, Moscow-Izhevsk, 2001.

[30] D. Bohm, *Quantum Theory*. Prentice-Hall, Inc., NY, 1952.

[31] M.B. Mensky, *Quantum Measurements and Decoherence*. Kluwer Academic Publ., 2000.

[32] D.I. Blokhintsev, *Principle Points of Quantum Mechanics (in Russian)*. Nauka publ., Moscow, 1966.

[33] M.A. Markov, *About Three Interpretations of Quantum Mechanics (in Russian)*. Librokom publ., Moscow, 2010.

[34] Avrelius Augustine, *Confession (in Russian)*. Azbuka, S.-Peterburg, 2008.

[35] Tit Lucretius Carus, *On the nature of things (in Russian)*. Hud. Lit. Publ. V. 2, Moscow, 1983.

[36] A. Einstein, *Radiation emission and absorption on the quantum theory, Collec. sci. papers. (in Russian)*. Nauka publ., V. 3, Moscow, 1966.

[37] A. Aspect, I. Dalibard, G. Roger, "Experimental Test of Bell's Inequalities Using Variable Analyzers," *Phys. Rev. Lett.,* vol. 49, pp. 1804-1807, 1982.

[38] B. Aspect. Bell's Theorem: The Naive View Of An Experimentalist. Quantum (Un)speakables - From Bell to Quantum information, R.A. Bertman, A. Zeilinger, Springer, 2002.

[39] A. Poincare, *Last thoughts (in Russian)*. Petrograd, 1923.

[40] G. Bateson, *Reason and Nature. Inevitable unity (in Russian)*. Librokom, Moscow, 2009.

[41] B. Russell, *Problems of philosophy (in Russian)*. S.-Peterburg, 1914.

[42] H. Bergson, Vocabulaire technique et critique de la philosophie, article «The unknowable». recited by A. Guter in his introduction to «Œuvres» by Bergson PUF, Paris, 1970.

[43] A.A. Grib, "Bell's Inequalities and Experimental Test of Quantum Correlations at macroscopic distances (in Russian)," *UFN.,* vol. 142, no.4, pp. 619-634, 1984.

[44] A. Einstein, V. Podolsky, N. Rosen, "Can quantum-mechanical description of physical reality be considered complete?" *Phys. Rev.,* vol. 47, pp. 777-780, 1935.

[45] J.S. Bell, "On the Einstein-Podolsky-Rosen Paradox," *Physics.,* vol.1, no.3, pp. 195-200, 1964.

[46] D. Bouwmeester, A. Ekert, A. Zeilinger, *The Physics of Quantum Information*. Springer, Berlin, 2000.

[47] J.A. Wheeler, R.P. Feynman, "Interaction with the Absorber as the Mechanism of Radiation," *Rev. Mod. Phys.,* vol. 17, pp. 157-181, 1945.

[48] J. Cramer, "Transactional Interpretation of Quantum Mechanics," *Rev. Mod. Phys.,* vol. 58, pp. 647-688, 1986.

[49] A. Fleming, C.L. Fortescue and oth, "Notes on modulation," *Nature.,* vol.125, pp.271- 306, 1930.

[50] A.S. Vinitsky, *Modulated filters and FM signals tracking reception (in Russian)*. Sov. radio, Moscow, 1969.

[51] M. Nielsen, I. Chuang, *Quantum computing and quantum information*. Cambridge University Press, 2000.

[52] D. Bouwmeester, J.-W. Pan, K. Mattle, M. Elbl, H. Weinfurter, A. Andzeilinger, "Experimental quantum teleportation," *Nature.,* vol. 390, pp. 575-579, 1997.

[53] P. Shor. Algorithms for quantum computation: discrete logarithms and factoring. In *Proceeding, 35-th Annual Symposium on Foundations of Computer Science, IEEE Press, Los Alamitos, CA, pages 124-134*, 1994.

[54] C.H. Bennett and G. Brassard. Quantum Cryptography: Public Key Distribution and Coin Tossing. In *Proc. IEEE Conf. on Computers, Systems, and Signal Processing, IEEE Press, 175-179*, 1984.

[55] J.I. Cirac, P. Zoller, "Quantum computations with cold trapped ions," *Phys. Rev. Lett.,* vol. 74, pp. 4091-4094, 1995.

[56] B.E. Kane, "A silicon-based nuclear spin quantum computer," *Nature.,* vol.393, pp. 133-137, 1998.

[57] D. Loss, D.P. DiVincenzo, "Quantum computation with quantum dots," *Phys. Rev. A.* vol. 57, pp. 120, 1998.

[58] R. Vrijen, E. Yablonovich, K. Wang, H.W. Jiang, A. Balandin, V. Roychowdhury, T. Mor, D.P. DiVincenzo. Electron Spin Resonance Transistors for Quantum computating *in Silico*n-Germanium Heterostructures. In *arXiv:quant-ph/990596 v2 11 Jun., pages 1-10*, 1999.

[59] Yu.A. Pashkin, T. Yamamoto, O. Astafiev, Y. Nakamura, D.V. Averin, J.S. Tsai, "Quantum oscillations in two coupled charge qubits," *Nature.,* vol. 421, pp. 823-826, 2003.

[60] D. Deutsch, "Quantum theory, the Church-Turing principle and the universal quantum computer," *Proc. Roy. Soc. Lond.,* vol. A 400, pp. 97-117, 1985.

[61] P.W. Shor, "Scheme for Reducing Decoherence in Quantum Computer Memory," *Phys. Rev. A,* vol. 52, pp. 2493-2496, 1995.

[62] A.M. Steane, "Error Correcting Codes in Quantum Theory," *Phys. Rev. Lett.,* vol. 77, pp. 793-797, 1996.

[63] A. Steane, *Quantum computing.* Reports Prog. Phys. 61, 117, 1998. Preprint, http://arxiv.org/archive/quant-ph/97080222 V 2, September 1997.

[64] L.K. Grover, "A Fast Quantum Algorithm for Database Search." *Proc. ACM Symp. on Theory of Computing, ACM Press, pages 212-219,* 1996.

[65] L.K. Grover, "Quantum Mechanics Helps in Searching for a Needle in a Haystack," *Phys. Rev. Lett.,* vol. 78, pp. 325-328, 1997.

[66] L.K. Grover, "A Framework for Fast Quantum Mechanical Algorithms," *Proc. ACM Symp. on Theory of Computing, ACM Press, pages 53-62,* 1998.

[67] P.W. Shor, "Polynomial-Time Algorithms for Prime Factorization and Discrete Logarithms on a Quantum Computer," *SIAM J. Computing.,* vol. 26, pp. 1484-1509, 1997.

[68] J.H. Poincare, *Selected Works V.3 (in Russian).* Nauka, Moscow, 1974.

[69] Chatterjee, Satischandra; Datta, DhirendramohanJ, *An Introduction to Indian Philosophy (in Russian).* Inost. Lit., Moscow, 2009.

[70] D. Sudzuki, *Bases of Zen-buddism (in Russian).* Odissey, Bishkek, 1993.

[71] V.M. Ames, *Zen and American thought.* University of Hawaii Press, Honolulu, 1962.

[72] red. V.V. Malyavin, B.B. Vinogradsky, *Antology of daoss philosophy (in Russian).* Tov. Klyshnikov-Komarov and K, 1994.

[73] M. Mesarovich, Bases of general theory of systems (in Russian). *The general theory of systems.* Mir, Moscow, 1966.

[74] A. Rosenblueth, N. Wiener, H. Bigelov, "Behavior, Purpose and Teleology," *Philos. Sci.,* vol. 11, pp. 18-24, 1943.

[75] R. Courant, H. Robbins, *What is Mathematics?.* Oxford University Press, Oxford, 1941.

[76] V. Boss, *Lectures in mathematics (in Russian), V.12.* KomKniga, 2008.

[77] L.P. Grimak, *Simulation of states rights in hypnosis (in Russian).* Nauka, Moscow, 1978.

[78] A.R. Luria, *Feelings and perceptions (in Russian)* Publ. of the Moscow University, Moscow, 1975.

[79] E. Zwicker, *Das Ohr als Nachrichtenempfanger.* S. Hirzel Verlag, Stuttgart, 1967.

[80] S. Buus, H. Musch, M. Florentine, "On loudness at threshold," *J. Acoust. Soc. Am.,* vol. 104, pp. 399-410, 1998.

[81] B. Scharf, J.C. Stevens, The form of the loudness function near threshold. *Proc. Int. Congress on Acoustics, 3rd, Stuttgard, 1959, edited by L. Cremer, Elsevier, Amsterdam, pages 80-82,* 1961.Sex and Gender Specific Medicine in Chronic Liver Diseases

INDEX

K

L

M

O

P

Q

R

S

www.ingramcontent.com/pod-product-compliance
Lightning Source LLC
Chambersburg PA
CBHW041717210326
41598CB00007B/682